Advances in Anatomy
Embryology and Cell Biology

Vol. 173

Springer
Berlin
Heidelberg
New York
Hong Kong
London
Milan
Paris
Tokyo

T. Heida

Electric Field-Induced Effects on Neuronal Cell Biology Accompanying Dielectrophoretic Trapping

With 46 Figures and 7 Tables

Springer

Dr. Tjitske Heida

University of Twente
Faculty of Electrical Engineering
Mathematics and Computer Science
Laboratory of Measurement and Instrumentation
Laboratory of Biomedical Engineering
P.O. Box 217
7500 AE Enschede
The Netherlands

e-mail: t.heida@el.utwente.nl

ISSN 0301-5556
ISBN 3-540-00637-0 Springer-Verlag Berlin Heidelberg New York

Library of Congress Cataloging-in-Publication Data

Heida, T. (Tjitske), 1972–
Electric field-induced effects on neuronal cell biology accompanying dielectrophoretic trapping / T. Heida
p. cm. – (Advances in anatomy, embryoloy, and cell biology,
ISSN 0301-5556 ; v. 173)
Includes bibliographical references and index.
ISBN 3-540-00637-0 (softcover : alk. paper)
1. Neurons. 2. Dielectrophoresis. 3. Nerves–Electric properties. 4. Microelectrodes. I. Title. II. Series.

Springer-Verlag a member of BertelsmannSpringer
Science+Business Media GmbH

http://www.springer.de

Typesetting: Stürtz, Würzburg

Printed on acid-free paper 27/3150Ag – 5 4 3 2 1 0

Contents

List of Symbols

α	Angle between the electric field line and a normal on the membrane [rad]
β	Fractional power
ϕ	Phase angle [rad]
∇	Vector differential operator $\nabla \equiv \vec{a}_x \frac{\partial}{\partial x} + \vec{a}_y \frac{\partial}{\partial y} + \vec{a}_z \frac{\partial}{\partial z}$
C	Capacity [F/m^2]
d_{cell}	Cell density [cells/ml]
δ	Membrane thickness [m]
Δ	Characteristic length over which the field varies [m]
$\varepsilon 0$	Permittivity of free space ε_0=8.85*10^{-12}F/m
ε	Relative permittivity [F/m]
ε^*	Complex permittivity $\varepsilon^* = \varepsilon - j\frac{\sigma}{\omega}$[F/m]
j	Imaginary unit j=(–1)$^{1/2}$
E	Electric field strength [V/m]
E_{rms}	Root mean square value of the electric field [V/m]
ψ	Volume fraction of cells
F	Force [N]
fCM	Clausius-Mosotti factor
f	Frequency [Hz]
ω	Radian frequency ω=2πf [rad/s]
K	Measure of magnitude
n	Number of experiments
r	Cell radius [m]
R	Resistance [Ω]
ρ	Resistivity [Ωm]
σ	Conductivity [S/m]
t	Time [s]
τ	Time constant [s]
T	Temperature [°C] or [K]
ΔT	Temperature difference [°C] or [K]
υ	Relaxation time [s]
V	Voltage potential [V]
V_m	Membrane potential [V]
V_{rest}	Resting membrane potential [V]
x, y, z	Cartesian coordinates
Z	Impedance [Ω]

1 Introduction

1.1 Neuro-Electronic Interfacing

1.1.1 Nervous System

Communication in the (human) body and the interaction with the environment is controlled by the nervous system. It can be divided into a central part, which includes the spinal cord, brainstem, cerebellum, and cerebrum, and a peripheral part, which includes all neuronal tissue outside the central part (Martini 2001). The latter provides the interface between the central nervous system and the internal and external environment of the body. Eye, ear, skin, and muscle sensors provide the necessary information. Via primary afferent neurons this information is transmitted to the central nervous system. Conversely, this system provides information to the motor organs via the efferent fibers. Furthermore, the central nervous system is responsible for cognition, learning, and memory.

Neurons are cells specialized for receiving information and transmitting signals to other neurons or to effector cells, such as muscles and glands (Levitan 1991). Like all other cells, neurons are enclosed by a cell membrane, which is a double layer of phospholipid molecules. This bilayer, about 10 nm thick, serves as a barrier that allows the cell to maintain an internal (cytoplasmic) composition far different from the composition of the extracellular fluid. It contains enzymes, receptors, and antigens that play central roles in the interaction of the cell with other cells. Many of the internal organelles are also enveloped in membranes, dividing the cell into discrete compartments and allowing the localization of particular biochemical processes in specific organelles. Many vital cellular processes take place in or on the membranes of the organelles.

Due to the large difference in composition in intra- and extracellular fluid, electrochemical potential gradients of certain ions exist across the plasma membrane. As illustrated in Fig. 1 these gradients result in the resting membrane potential, which is generally about −70 mV. Voltage-gated ion channels in the cell membrane of neuronal cells are able to generate rapid changes in the membrane potential by ionic currents. When the membrane potential is depolarized beyond a critical threshold value (i.e., −55 mV) an action potential is created. The membrane potential becomes positive within about a millisecond and attains a value of about +30 mV before turning to the negative resting membrane potential again. These action potentials form the basis of the signal-carrying ability of neuronal cells. Intracellular information trans-

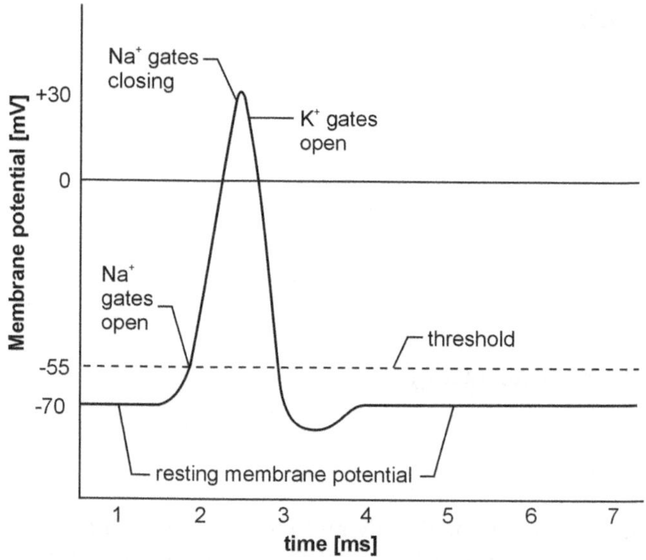

Fig. 1 Action potential. Exceeding the threshold potential results in an Na^+ influx. At about 30 mV the Na^+ gates close and an K^+ efflux starts bringing the membrane potential back to the resting membrane potential (–70 mV). These rapid changes occur within a time period of about 5 ms

fer is realized by the axon, which is a thin process arising from the neuronal cell body. Dendrites are also neuronal processes, but thicker and shorter than axons and they are often highly branched. They form the sites at which information is received from other neurons. Finally, the synapse, the most highly specialized structure, is the point where information transfer between neurons takes place via neurotransmitters.

The environment of most neurons is controlled so that they are normally protected from extreme variations in the composition of the extracellular fluid. Neuroglia help maintain an appropriate local environment for neurons. Schwann cells, a specified type of glia cells, are responsible for the creation of the myelin sheath around axons in the peripheral nervous system, while centrally the oligodendrocytes fulfill this task. This sheath increases the speed of propagation of action potentials and allows action potentials to be generated only at the nodes of Ranvier (the gaps between adjacent Schwann cells or oligodendrocytes).

1.1.2 Restoring Neuronal Functions

Severe injury to nervous tissue causes cell death. Neurons can hardly be replaced, since they are postmitotic cells. Thus, with the loss of a neuron also the synaptic connection with other neurons is lost, and no information can be transmitted through the disconnected network anymore. Muscle activation or sensory functions like hearing and vision may be impaired. However, the anatomic structure of the disconnected network driving these functions may still be intact and can therefore still be activated, but the stimulation needs to come from an artificially generated (external) signal. For example, electrodes placed on the skin under which a particular muscle is located can be used to activate this muscle directly. However, for accurate control specific muscle fibers need to be activated without activating other muscles. This also applies to the stimulation of nerve fibers. Therefore, the ideal neuro-electronic interface is fiber selective. Fiber selectivity can only be realized by the creation of a very close contact between electrodes and nerve fibers. Nerve cuff techniques have been introduced for selective activation of peripheral regions of a nerve trunk, containing multiple fascicles (Sweeney et al. 1990). Additionally, arrays consisting of many electrodes [e.g., 32 rows of each 4 electrodes arranged in a rectangular grid (Smit et al. 1999)] have been developed to increase selectivity. These arrays were inserted into a rat peripheral nerve in such a way that each electrode could stimulate one or a few axons. The efficiency of these arrays, defined as the fraction of electrodes selectively contacting a motor unit and evoking a corresponding threshold force, was about 80% (Smit et al. 1999). By using different heights for the electrodes a three-dimensional configuration is created allowing the stimulation of nerve fibers that are located in different fascicles of the nerve. In spite of a high selectivity, practical use of these three-dimensional electrode arrays is limited by the complex fabrication procedure. Furthermore, determination of the correct position for implantation is difficult. Therefore, a new concept was developed: the cultured neuron probe. This is a planar electrode array that can be fabricated using "standard" photolithographic and etching techniques.

An implanted microelectrode array (MEA) on top of which groups of neuronal cells are cultured creates an optimal environment for the connection with the neuronal tissue of the human nervous system. Collateral sprouts incited to branch from a motor fiber can be guided towards and contacted with a group of cultured cells thereby creating a very selective and efficient interface (Rutten et al. 2001). Stimulation of this system will then result in an action potential that propagates via the collateral sprout to the neurons in the body in order to reach the muscle to be activated.

1.2 Culturing Neuronal Cells

1.2.1 Dissociation

Neuronal cultures are preferably prepared from embryonic or early postnatal animals (Banker and Goslim 1998). One of the reasons for this preference is that after dissecting a small part of tissue (e.g., from the cerebral cortex) it must be dissociated to give a suspension of single cells. The stronger the cell-cell associations, such as found in synaptic junctions, the more difficult it is to dissociate tissue. Mature tissue is thus difficult to dissociate without damaging its constituent cells.

Protease treatment is frequently used for dissociation. Tissue to be dissociated is cut into small fragments and is incubated with protease at 37°C for a certain time period (several minutes to an hour or more). Enzymes are large molecules that do not enter cells and are therefore suitable to remove externally exposed cell surface proteins and proteins of the extracellular matrix without badly damaging the cell

(Banker and Goslim 1998). However, during this process other cell surface components may also be attacked; e.g., trypsin treatment preferentially destroys some classes of transmitter receptors that reappear after about a day in culture.

After enzyme treatment, protease activity must be stopped. This is accomplished by rinsing the tissue or removing the fluid above the pellet of cells after centrifugation. Other methods include the addition of serum or of a specific inhibitor, e.g., soybean trypsin inhibitor. Afterwards, the remaining tissue fragments can be dissociated mechanically by trituration (the shear forces created by sucking up the cell suspension in a pipette and releasing it afterwards).

1.2.2 Culturing Conditions

At all stages, tissues and cells must be maintained in an osmotically balanced solution at physiological pH (pH~7). A balanced salt solution (BSS) consists of a mixture of salts (including Na, K, Mg, Ca, Cl, PO_4, and HCO_3) at concentrations approximating those of extracellular fluid, together with glucose. Culture medium consists of such a basal medium supplemented with nutrients needed for long-term survival and growth of cells (amino acids, vitamins, and other ingredients) and serum or, in the case of a chemically defined medium, a defined set of hormones and growth factors (Romijn et al. 1984; Marani et al. 1988). Buffers are used for equilibration of the medium in a certain atmosphere (e.g., 5%CO_2, 37°C, pH~7). For equilibration in air, a combination of low concentrations of $NaHCO_3$ and PO_4 or a buffer, such as HEPES, is used (Banker and Goslim 1998).

Cultures are maintained in an incubator perfused automatically with a mixture of air and CO_2, depending on the medium chosen (Banker and Goslim 1998). To minimize evaporation of the medium high humidity is necessary.

In vitro adhesion of cells involves cell-cell and cell-matrix interactions mediated by specific cell surface receptors (Banker and Goslim 1998). These interactions can be mimicked in culture when extracellular matrix molecules or cell adhesion molecules are used as a substrate. Glass itself is negatively charged, and since the net charge on cell surfaces is also negative, it is therefore thought to be a poor substrate. For most purposes the glass (or plastic) substrate is treated or chemically modified to improve its adhesive properties. Adsorption of macromolecules or pretreatment with polymers of basic amino acids [e.g., polylysine or polyethylenimine (PEI)] may be beneficial (Rüegg and Hefti 1984). Polylysine also enhances adsorption of other substrate proteins, such as extracellular matrix constituents (Banker and Goslim 1998).

In general, cell attachment is complete within a few hours of plating. Once attached, the number of surviving neurons should remain constant for an extended period (up to about a month, and sometimes even several months), with loss of only a small percentage of cells per week. However, at some point neurons throughout the entire culture start to degenerate, heralding the end of the culture. Early signs of degeneration are cell body swelling and beading of processes.

It was found that neurons in culture show all the morphological features seen in cortical neurons in vitro (Kriegstein and Dichter 1983). This was investigated by

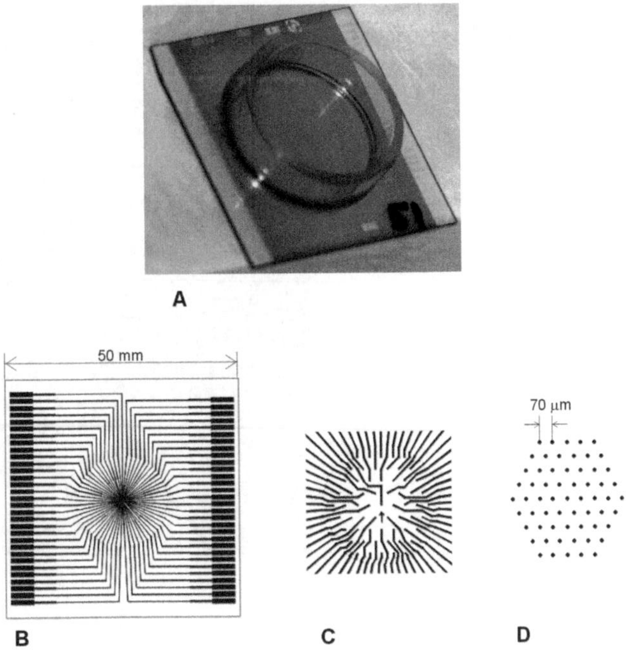

Fig. 2 A Multielectrode array (50×50 mm) for culturing neurons, **B** complete layout, **C** inner electrode structure in detail, **D** positions of the electrode tips with interelectrode distances of 70 μm

characterizing cell form (pyramidal, multipolar, or fusiform). Additional features of neuronal cells are a large, clear nucleus with a prominent nucleolus, dendritic branching pattern, spine density, and axonal projections.

1.3
Positioning and Culturing Neuronal Cells on a Microelectrode Array

1.3.1
Microelectrode Array

Given the current possibilities in the field of microtechnology, new research methods in the biological sciences have been developed. An important tool that has been used since about 1972 is the microelectrode array (MEA). Initial experiments with it were focused on recording the spontaneous activity of cardiac cells (Thomas et al. 1972). Later on it became a tool in neurotechnology for recording and stimulation purposes (Gross 1979; Pine 1980).

Up till now the microelectrode array used for stimulation and recording experiments with neuronal cells and tissue at the University of Twente has consisted of a glass substrate (5 cm×5 cm) on top of which an array of electrodes is created by lithographic and etching methods. Figure 2 shows the layout of the MEA. Around the center a glass ring is mounted to provide a culture chamber for the neuron suspension. The electrodes are insulated except for a small hole on top of the electrode tip.

Fig. 3 Cultured dorsal root ganglion (DRG) cells from neonatal rats randomly positioned on top of an MEA at 1 day in vitro (1 DIV)

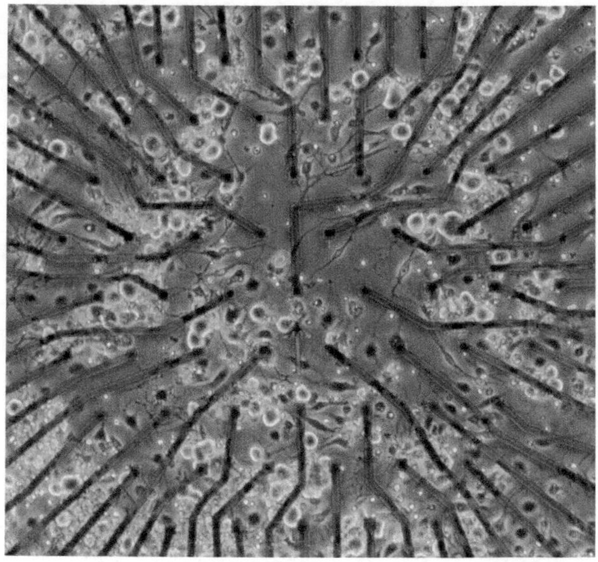

1.3.2
Cell Positioning

Applying a neuron suspension onto these MEAs will result in a random positioning of the neurons and an uncontrolled outgrowth as shown in Fig. 3. However, the aim of using the MEA for selective stimulation requires that separate groups of neurons are positioned on top of each of the electrode sites. By using different kinds of manipulation techniques it is possible to control the positioning and outgrowth of neurons.

Chemical modification of the substrate can be used to make certain regions of the surface more or less attractive for the neurons to adhere to, i.e., neurophilic and neurophobic substrates, respectively (Kleinfeld et al. 1988; Corey et al. 1991, 1996; Matsuzawa et al. 1993, 1996; Curtis et al. 1995; Branch et al. 1998; Rutten et al. 2001). For example, fluorocarbon and polytetrafluoroethylene may be used to provide nonadhesive layers, while polyethylenimine creates a nonspecific neuron-adhesive substrate (Rüegg and Hefti 1984; Ruardij et al. 2000). By using chemical modification in different patterns, it may be possible to position the neurons on particular spots and even to define the paths for dendrites and axons to follow.

Mechanical modification may contribute to this method since both dendritic and axonal outgrowth follow deep grooves in the substrate (Clark 1987, 1991a, b; Wilkinson 1995). In making use of adhesiveness, a drop of cell suspension is applied atop the substrate, and the cells will precipitate randomly. Cells that are not located within the neurophilic regions are generally lost.

Another method of trapping cells is to make use of dielectrophoresis (DEP). With dielectrophoretic forces, cells can be confined to a certain area without loss.

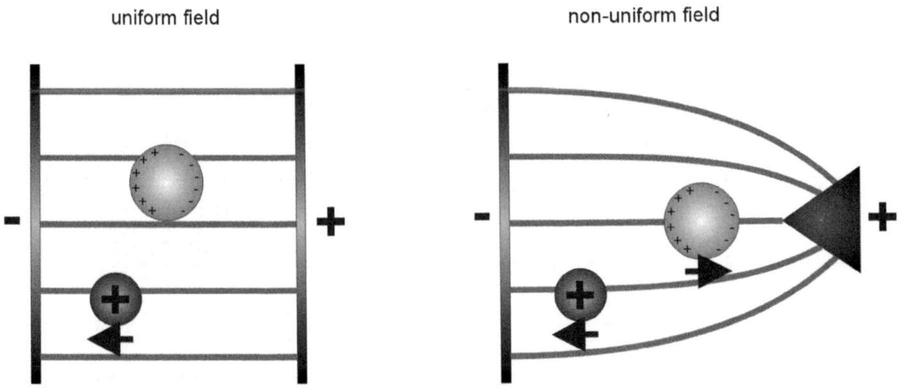

Fig. 4 The principle of dielectrophoresis. A (positively) charged and a neutral particle located in uniform and nonuniform electric fields. A translational force is induced on the neutral particle only when it is exposed to the nonuniform field, while the charged particle is directed towards the electrode of opposite charge in both cases

1.4 Dielectrophoresis

1.4.1 Principle of Dielectrophoresis

Any electrical field, uniform or nonuniform, exerts a force upon a charged body (electrophoresis), but it is characteristic of nonuniform electric fields that they exert a force upon neutral bodies. Dielectrophoresis (DEP) is the translational motion of neutral matter caused by polarization effects in a nonuniform electric field (Crane and Pohl 1972; Pohl 1978; Pethig et al. 1992; Foster et al. 1992; Schnelle et al. 1993, 1996; Wang et al. 1993; Markx et al. 1995, 1996; Fuhr et al. 1992, 1994a, b, c, 1995; Müller et al. 1996, 1999; Jones and Washizu 1996; Green et al. 1997). The resulting force is termed the dielectrophoretic force.

Figure 4 shows an illustration of a uniform and a nonuniform electric field in which a charged as well as a neutral body reside. In the uniform field, the charged particle (in this case positively charged) is pulled along the field lines toward the electrode of opposite charge. In the same field, a neutral body will polarize: a negative charge is created on the side near the positive electrode, while a positive charge is created on that near the negative electrode. The net result may be a torque (e.g., if the body is anisotropic), but no net translational force, without which the body as a whole will not move toward either electrode.

In the nonuniform field, a charged object is still attracted toward the electrode of opposite polarity. In this case a translational force is also exerted on the neutral particle. Under the influence of the field it acquires a polarization. Since the particle is neutral, the two charges on the body are in fact equal, but the fields operating on each side are not equal. The neutral body is, in the situation sketched in Fig. 4, directed towards the region of high field strength by a positive dielectrophoretic force. No matter which electrode is charged positive or negative, the force is always in the

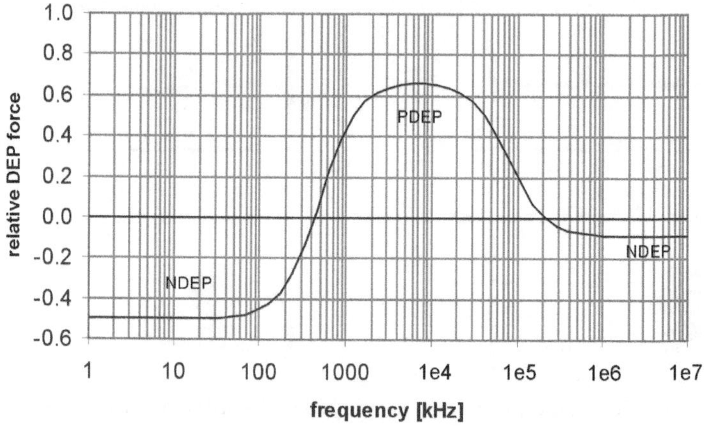

Fig. 5 Relative dielectrophoretic force, determined by the real part of the Clausius–Mosotti factor, as a function of frequency for a simple homogeneous sphere with a thin membrane. Frequency ranges over which negative (*NDEP*) and positive (*PDEP*) dielectrophoretic forces occur are indicated

same direction. Therefore, the applied field could well be that due to an alternating potential since the polarization can usually reverse as the field switches.

The direction of the dielectrophoretic force depends on the electrical properties of the particle and the suspending medium. The Clausius-Mosotti factor, f_{CM}, describes the interrelationship of these frequency-dependent properties of both materials. When the polarizability of the medium exceeds that of the cells, negative dielectrophoretic forces (NDEP) are created on the cells, which means that they are repelled from the electrodes (the regions of high field intensities). Otherwise the cells are attracted to the electrodes (positive dielectrophoretic forces—PDEP). The actual shape of the curve relating frequency and dielectrophoretic response, the spectrum, varies with the type of organism and with its physiological state. An example of a typical spectrum is shown in Fig. 5. With moderate medium conductivities (around 0.1 S/m) there are distinct frequency regions in which NDEP or PDEP occurs. This property makes dielectrophoresis an excellent tool for separating cells on the basis of differing electrical properties. Examples are the separation of bacteria from red blood cells (Wang et al. 1993), viable from nonviable yeast cells (Wang et al. 1993) and normal from cancerous cells (Gascoyne et al. 1993).

Previous research in the field of dielectrophoresis (DEP) has already shown that small particles (such as viruses and bacteria) and living cells (e.g., yeast cells, suspensor protoplasts, 3T3 fibroblasts) can be trapped by dielectrophoretic forces (Pethig et al. 1992; Fuhr et al. 1994c, 1995; Markx et al. 1995, 1996). Field strengths of two to several hundred kilovolts per meter, which can easily be created by microelectrode structures, are required for trapping (Fuhr and Shirley 1995; Green and Morgan 1997; Morgan et al. 1999). Arrays of planar electrodes positioned in parallel, quadrupole, and octopole configurations, as well as three-dimensional field cages, were found to position or trap cells efficiently. Figure 6 shows an example of a quadrupole field trap for the collection of viruses (Schnelle et al. 1996).

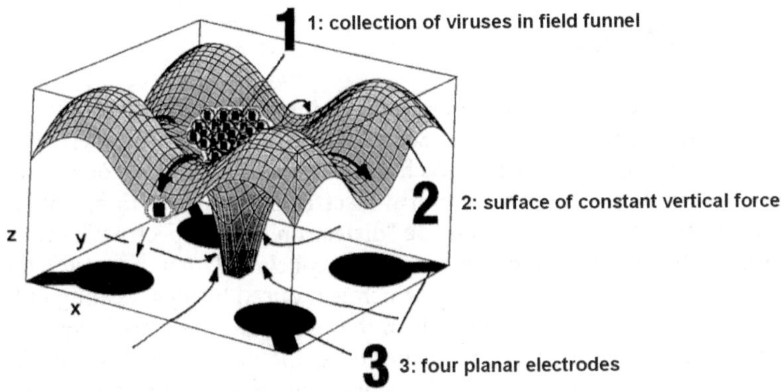

Fig. 6 Field "funnel" created by a planar quadrupole electric field in which particles (viruses) are trapped by negative dielectrophoretic forces. (From Schnelle et al. 1996)

1.4.2 Viability of Cells Exposed to Electric Fields

A very important aspect of the use of dielectrophoretic forces for trapping neuronal cells on the electrodes of an MEA is the viability of the cells after being exposed to the electric field. Due to the capacitive behavior of the membrane, an electric field induces a membrane potential "superimposed" on the resting membrane potential. The capacitive representation, however, fails when large membrane potentials are induced by high electric field strengths; the membrane potential settles into a nearly constant value. This value was found to be of the order of 0.2–1 V for several cell types, but not including neuronal cells (Benz 1980a, b, c; Teissié and Rols 1993; Zimmermann and Neil 1996). At this point membrane breakdown occurs, which is also termed electropermeabilization or electroporation. The membrane acquires a very large conductance assigned to the creation of electropores, i.e., conducting aqueous pathways, in the membrane. Through these pores ions and small molecules may be transported (Weaver 1993).

The breakdown of a cell membrane may be a reversible process, i.e., the original low conductivity and permeability state may be restored within several minutes. However, the return to a low conductivity and permeability state does not automatically imply that the original membrane structure is completely restored and that the cell behaves like it would without being exposed to the electric field. Cell death may occur due to rupture of the membrane or chemical imbalances resulting from influx and efflux during the high-permeability state of the membrane.

Several types of living cells proved to be capable of surviving continuous exposure to rather high electric fields for time periods up to two days. This was shown for red blood cells (Bao et al. 1992), mouse fibroblasts (3T3, L929; Fuhr et al. 1994a), suspensor protoplast (Fuhr et al. 1994b), bacteria, and yeast (Crane and Pohl 1968; Mason and Townsley 1971; Markx and Pethig 1995).

1.5
Scope of This Review

It has been shown that cortical rat neurons can be trapped effectively by negative dielectrophoretic forces, as outlined in Sect. 2. Simple theoretical models were used to describe the trapping process for which a quadrupole microelectrode structure was used for the creation of the nonuniform electric field. In addition, several experiments were performed to investigate the "dielectrophoretic spectrum"; the results are described in the final part of Sect. 2. It is concluded that not only the dielectrophoretic force but also a fluid flow, which is induced by local heating of the medium, is responsible for trapping the cells in the center of the electrode structure, especially at frequencies above 1 MHz. Furthermore, the field strength is not directly related to the amplitude of the sinusoidal input signal since the electrode-medium interface creates a large impedance, especially at low frequencies, which reduces the field strength inside the medium.

As described in a previous section, dielectrophoretic trapping of particles of micrometer dimensions requires rather high field strengths, which may induce large membrane potentials, possibly resulting in electroporation of the membrane. Therefore, a considerable part of the investigation described in this review focuses on the effects of exposing neuronal cells to nonuniform electric fields.

In Sect. 3, a detailed theoretical description of membrane breakdown is given. However, most experiments on membrane breakdown are generally focused on cells exposed to uniform electric fields. Moreover, according to literature, the effect of nonuniform fields on electroporation is considered to be unpredictable. Especially, it is difficult to determine which part of the membrane is affected by electroporation without sophisticated observational techniques. Therefore, the distribution of the membrane potential across the cell membrane for a neuronal cell positioned in a nonuniform electric field is predicted by using a finite element method. In addition, experiments were performed to investigate the membrane breakdown level for neuronal cells for frequencies up to 1 MHz.

Section 4 describes the investigation of the conditions under which preservation of cell viability is expected (theoretically), i.e., low field strengths (<80 kV/m) and a high frequency (14 MHz). Cell morphology (cell growth, and the number and length of their processes), as well as the number of outgrowing and nonoutgrowing cells as compared to nonexposed cells, were the parameters that were evaluated. Additionally, cells trapped at lower frequencies and higher amplitudes were tested on viability using a staining method. Outer appearances of the neuronal cell, however, are not decisive for their functional behavior. Therefore, a pilot study investigating spontaneous activity of cortical rat neuronal cells after being trapped dielectrophoretically is described at the end of Sect. 4. Finally, Sect. 5 gives a summary.

2 Dielectrophoretic Trapping of Neuronal Cells

In this section, it is shown that dielectrophoresis can effectively be used to trap cortical rat neurons. The trapping process was found to depend on amplitude and frequency in a nonlinear way. Several aspects were theoretically and experimentally investigated in order to come to a better understanding of the trapping process.

2.1 Theory

2.1.1 Dielectrophoretic Force

The time-averaged dielectrophoretic force F_{DEP} exerted on a spherical particle (indicated by "p," suspended in a medium, "med") exposed to an alternating current (AC) electric field is dependent on the in-phase component of the dipole moment, and can be written as:

$$\vec{F}_{DEP}(\omega,x,y,z) = 2\pi r^3 \varepsilon_{med} \mathrm{Re}\left[\frac{\varepsilon_p^*(\omega) - \varepsilon_{med}^*(\omega)}{\varepsilon_p^*(\omega) + 2\varepsilon_{med}^*(\omega)}\right] \nabla \vec{E}_{rms}^2(x,y,z) \quad (1)$$

where the index rms denotes the root-mean-square value of the electric field, $\omega=2\pi f$ with f the frequency of the electric field, r is the radius of the particle, and ε^* is the complex permittivity $\varepsilon^*(\omega) = \varepsilon - j\frac{\sigma}{\omega}$, where $j=(-1)^{1/2}$, ε and σ are the permittivity and conductivity, respectively, ∇ is the vector differential operator $\nabla \equiv \vec{a}_x \frac{\partial}{\partial x} + \vec{a}_y \frac{\partial}{\partial y} + \vec{a}_z \frac{\partial}{\partial z}$.

According to Eq. 1, the DEP force is a function of the dimensions of the particle, the electrical properties of the particle and the medium, and the distribution of the electric field. The electrical properties of particle and medium are included in the so-called *Clausius-Mosotti factor* (f_{CM}).

$$f_{CM}(\omega) = \frac{\varepsilon_p^*(\omega) - \varepsilon_{med}^*(\omega)}{\varepsilon_p^*(\omega) + 2\varepsilon_{med}^*(\omega)} \quad (2)$$

The real part of this factor gives the frequency dependence and the direction of the DEP force (Eq. 1). When the conductivity and/or permittivity of the medium, each dominating in the kHz- and MHz-range, respectively, exceeds that of the cell interior, negative DEP forces will occur $\left(\mathrm{NDEP}: \left|\varepsilon_p^*\right| - \left|\varepsilon_{med}^*\right| \langle 0\right)$.

The particles are then repelled from regions of high field intensities. Otherwise, positive forces lead to attraction $\left(\text{PDEP}: \left|\varepsilon_p^*\right| - \left|\varepsilon_{med}^*\right| \rangle 0\right)$.

The derivation of Eq. 1 is based on the following assumptions (Pohl 1978; Jones and Washizu 1996; Crane and Pohl 1972; Washizu 1992; Wang et al. 1994):

- The applied field varies sinusoidally in time.
- The media are isotropic and linear, i.e., ε and σ are not functions of the field orientation or the field strength.
- The frequency of the field is low enough to permit the electric field to be derived from a scalar potential.
- The external field is assumed to be uniform in the vicinity of the sphere.
- A single spherical particle is considered.

2.1.2
Electrical Properties of Cells and Cell Suspensions

The active basic molecular mechanisms responding to an external field are described as polarization and conduction (Pohl 1978). Freely moving charges represent the conductivity, while their response appears as a polarization if their motion is blocked. Real dielectric materials exhibit both polarization and conduction and linked together these properties form the complex permittivity of the materials. Several modes of polarization exist, each having its own relaxation mechanism and relaxation time, i.e., at a certain frequency of the external field, polarizability is reduced since the dipoles are not able to follow the field anymore; this is termed dispersion. Four dispersions can be distinguished in the range up to about 100 GHz, namely, α-, β-, δ-, and γ-dispersion (Saito et al. 1966: Gimsa and Wachner 1998). The α-dispersion appears at frequencies up to 1 kHz, the β-dispersion ranges from 1 kHz to 100 MHz, the δ-dispersion appears in the range of 1 to 10 GHz, and the γ-dispersion appears at frequencies over 10 GHz. These frequency ranges are not very strictly defined and several different subdivisions can be found in literature (Kaler and Jones 1990; Pethig 1990; Gimsa and Wachner 1998).

Several mechanisms are proposed to contribute to the α-dispersion, including charging of intracellular structures connecting with the outer membrane, and frequency dependencies of the membrane itself (Pethig 1990). Another mechanism lies within the ionic atmosphere. Cells and microorganisms carry a net negative surface charge of around 10 mC/m^2. Counter-ions will be attracted and will create an electrical double layer around the cell, which can be polarized by an external field. The movement of (bound) ions along the cell surface due to an external field corresponds to a high local conductivity. A potential change is produced which induces a displacement of the nearby counter-ion atmosphere, the polarization current. Since the movement of the ions along the surface of the particle is relatively massive and slow, this type of polarization is effective only at relatively low frequencies.

The β-dispersion is generally assigned to the Maxwell–Wagner interfacial polarization mechanism, which is considered as the main component responsible for dispersions in biological cells (Wang et al. 1993; Gimsa and Wachner 1998). Interfacial polarization is a consequence of different dielectric constants and conductivities of two media (e.g., particle and suspending medium). It appears around 1 MHz for cell suspensions and biological tissues (Pethig 1990; Asencor et al. 1990).

The δ-dispersion is assigned to small molecules or bound water, while the γ-dispersion is assigned to the relaxation of free water (Gimsa and Wachner 1998). The frequency ranges of the δ- and γ-dispersion normally exceed the experimental frequency range used for dielectrophoretic manipulation of cells.

2.1.3 Modeling the Electrical Properties of a Suspended Biological Cell

Back in 1891, Maxwell developed an equation describing the complex permittivity of a suspension containing a certain volume fraction of suspended particles. This is the "mixture equation," of fundamental importance in defining the dielectric properties of heterogeneous systems in general. It was refined to describe multi-shell models which could represent heterogeneous particles such as biological cells (Irimajiri 1979). Several shells that envelop a homogeneous "nucleus" can represent the internal cell contents. At each interface there will be a build-up of charge at a characteristic relaxation rate, giving rise to a Maxwell–Wagner dispersion. It was demonstrated that such a concentric system could be replaced by a homogeneous sphere of the same outer radius having an "effective complex permittivity," which means that when placed in an electrical field the homogeneous sphere could be substituted for the heterogeneous sphere without altering the field.

Applying multi-shell models requires knowledge of the electrical properties of all of the compartments (Crane and Pohl 1972; Drago and Ridella 1982; Fuhr et al. 1995a; Hughes et al. 1998; Irimajiri 1979; Marha and Musil 1975; Pastushenko et al. 1985; Pohl 1978). However, in principle a cell can be seen as consisting of an interior composed of a rather well conducting saline fluidic material, around which a membrane is present, i.e., it can be represented by a single-shell model (Pohl 1978; Irimajiri 1987; Zieliński 1989; Gimsa et al. 1991, 1998; Gascoyne et al. 1993; Kaler and Jones 1990). Figure 7 shows the principle of using an effective complex permittivity for the single-shell model.

The single-shell model can give rise to one (intrinsic Maxwell–Wagner, or β-) dispersion (Irimajiri 1979; Kaler and Jones 1990; Gimsa et al. 1991; Foster et al. 1992; Wang et al. 1992, 1993). Using the fundamental equation defining the dielectric prop-

Fig. 7 Single-shell model. Combination of the electrical properties of the membrane and cytoplasm into one effective complex permittivity; σ_{med}, and ε_{med} are the conductivity and permittivity of the medium, respectively; σ_{int}, and ε_{int} are the conductivity and permittivity of the cytoplasm, respectively, ε_{mem} is the permittivity of the membrane, δ is the membrane thickness, r is the cell radius and ε^*_{eff} is the effective complex permittivity of the cell

erties of heterogeneous systems in general, the effective complex permittivity for the single-shell model can be described in terms of the permittivities and conductivities of the two regions. Since the thickness of the membrane δ is small compared to the radius r of the cell (δ/r<<1), this equation can be written as (Irimajiri 1979; Kaler and Jones 1990):

$$\varepsilon^*_{ss,eff} \approx C_{mem} r \left[\frac{j\omega\tau_{\text{int}}+1}{j\omega(\tau_{mem}+\tau_{\text{int}})+1}\right] \tag{3}$$

where $\tau_{\text{int}}=\varepsilon_{\text{int}}/\sigma_{\text{int}}$ and $\tau_{\text{mem}}=\varepsilon_{\text{mem}}r/\sigma_{\text{int}}$ are time constants, C_{mem} is the membrane capacitance per unit area, $C_{\text{mem}}=\varepsilon_{\text{mem}}/\delta$, with the index ss denoting the single-shell model, "int'"the inner sphere, and "mem" the membrane.

The membrane, having a thickness in the order of 10 nm, has a lipid layer bounded on both sides by polar material. Electrically, the membrane can be represented by a capacitance in parallel to a low conductance. The capacitance of a membrane is about 1 mF/m^2 and its conductance ranges from 10^{-7} to 10^{-4} S/m (Saito et al. 1966; Pohl 1978; Pethig 1990). Water or ionic solutions have a relaxation frequency above 10^{10} Hz. This means that up to this frequency the permittivity and conductivity of the interior can be considered real and independent of frequency.

For a suspended cell a third region is present, the suspending medium which, due to the interface between cell and medium, adds another Maxwell–Wagner dispersion. By the same argument as that for the cell interior, the permittivity and conductivity of the suspending physiological medium (essentially a diluted solution of salt ions in water) can be considered real.

The effective complex permittivity of the cell together with the complex permittivity of the medium determine the Clausius–Mosotti factor. In effect, this factor may now be written as:

$$f_{CM} = f_{CM,\infty} + \sum_{k=1}^{2} \frac{\Delta f_{CM,k}}{1+j\omega\upsilon_k} \tag{4}$$

with $\Delta f_{CM,k} = 2\pi\varepsilon_m r^3 \left|\nabla E^2_{0,RMS}\right| \Delta f_k$(k=1, 2), the dispersions in the Clausius–Mosotti factor, with relaxation times v_1 and v_2, $f_{CM,\infty}$, is the limiting high-frequency value of f_{CM}. The relaxation times correspond to frequencies coinciding with the maximum rate of change of the DEP force.

2.2 Materials

2.2.1 Planar Quadrupole Microelectrode Structure

Figure 8 schematically shows the quadrupole microelectrode structure that was used for the experiments described in this review. Two types of electrodes are shown, with triangular tips and semi-circular tips, and both with an inter-electrode distance between the diagonally opposing tips of 100 μm. The latter was only used for the investigation of cell viability after trapping, as described in Sect. 4. Four structures of each

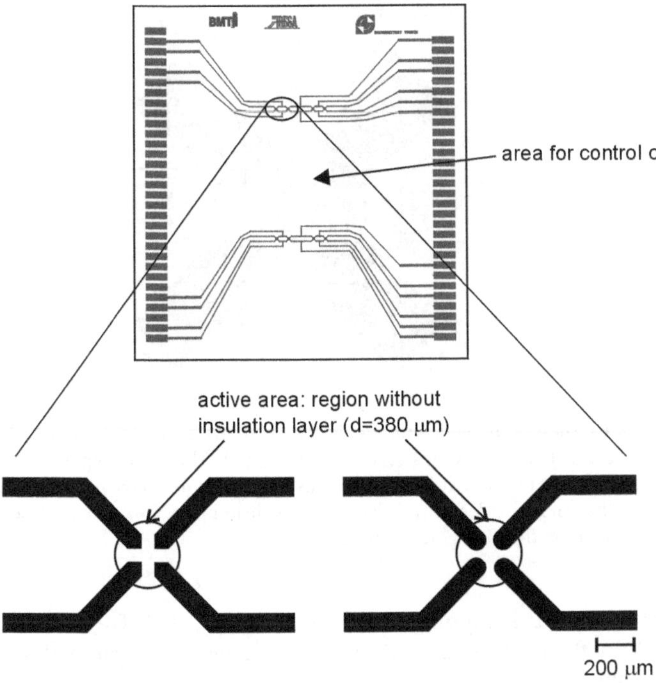

Fig. 8 Electrode plate with several quadrupole electrode structures that are schematically drawn in the *lower part* of the figure. Two types of electrode tips were used: triangular and semicircular shaped tips. The diagonal inter-electrode distance is 100 μm for both configurations. The central part of the electrode plate was used for reference cells, i.e., cells not exposed to electric fields but cultured under the same conditions as the trapped cells

type were fabricated on the same electrode plate with connection sites located at two sides of the plate for the connection with electronic devices.

A glass substrate was used because of its transparency. On the substrate of 5 by 5 cm (thickness 1 mm) titanium was applied to improve the adherence of a gold layer to the glass. Gold electrode structures were created by photolithography and reactive ion etching (RIE). An insulation layer [consisting of a silicone nitride sheet (Si_xN_y) of 396 nm sandwiched between two layers of silicon oxide (SiO_2) of 144 nm each] was set in place by plasma-enhanced chemical vapor deposition (PECVD). This insulation layer was then etched away at the "active regions" (the circular regions centered around the electrode tips) and at the connection sites located at two edges of the glass plate. After fabrication the electrode plates were cleaned in an ultrasonic bath with acetone. For the application of a neuron suspension, a culture chamber (a glass ring with an inner diameter of 3 cm) was mounted onto the electrode plate, comparable to the MEA as described in Sect. 1.

A more detailed drawing of the quadrupole structure is given in Fig. 9. In this figure the different areas that were used for the analysis of the experimental data are indicated. For most experiments the center was the area under investigation. In this area the cells are trapped by negative dielectrophoretic forces and it is in this area

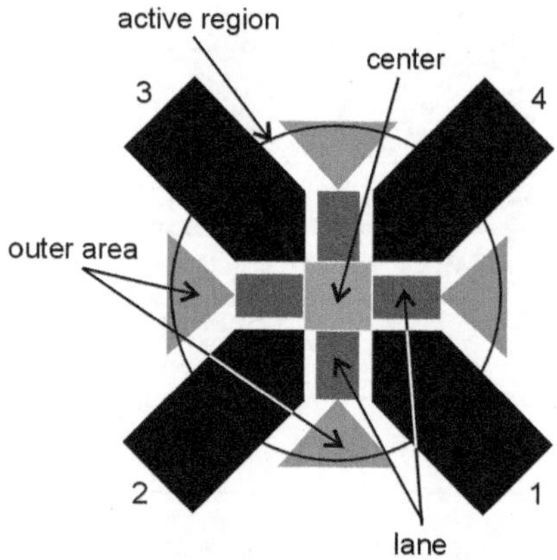

Fig. 9 The electrode structure in detail with the definition of the areas: The center is the square area that is formed by the four electrode tips (center = 5×10^{-9} m^2). The active region is the region centered around the electrode tips which is free from insulation. Its diameter is 380 μm. The lanes are the areas between the parallel sides of the electrode tips leading to the center. The outer areas are triangular areas formed by the diverging electrodes

that cell viability after the trapping process was investigated. The lanes and outer areas were used for the investigation of the trapping process itself as described in the following sections.

2.2.2 Electric Field Generation

For the generation of the electric field in the kHz range and MHz range, the following signal generators were used, respectively:

- HP 3310 (Hewlett-Packard, Inc., Palo Alto, CA, USA). A signal generator with frequency range 0.0001 Hz–5 MHz, amplitude range 1 V–16 V
- Tektronix Type 191 (Tektronix, Inc., Beaverton, OR, USA). A signal generator with frequency range 350 kHz–100 MHz ±2%, amplitude range 5 mV—5.5 V ±5%

For the investigation of the membrane breakdown potential of cortical rat cells as described in Sect. 3, a programmable signal generator was used; a HP 3245 A universal source. Amplitude and frequency were set via a computer according to a Labview script (LabVIEW 6.0, National Instruments LabVIEW, 2000, Austin, TX, USA) in the range from 2 to 18 V_{pp}, and 100 kHz to 1 MHz, respectively.

2.2.3
Cells and Medium

2.2.3.1
First Series

Cortical neurons from rat fetuses of embryonic day 18 (E18) were used for the initial trapping experiments, as described in the following section. The brains were taken out after decapitation, the meninges of the cortices were removed and the basal ganglia as well as the hippocampus were prepared free. The remaining cortices were collected in a tube filled with chemically defined R12 culture medium (Romijn et al. 1984; Marani et al. 1988) and mechanically dissociated. After precipitating undissociated parts, the upper layer of the suspension was removed and centrifuged at a rate of 1,200 rpm for 5 min. The medium above the centrifuged neurons was removed and the neurons left at the bottom were again mechanically dissociated using a drop of fresh medium. The last two steps were repeated once. A dilution of 10^6 neurons/ml was used.

2.2.3.2
Second Series

Cells obtained exclusively from the telencephalic cortex of newborn rats at postnatal day 2 (P2) were used for the investigation of cell viability after trapping. The brains were taken out after decapitation, and the cortices of the forebrain were carefully collected in a Petri dish and cut into small pieces. A drop of chemically defined R12 culture medium (Romijn et al. 1984; Marani et al. 1988) was added and the cortex parts with the medium were placed in a conical tube (during the dissociation procedure no glutamine, vitamin C, or nerve growth factor (NGF) were added to the R12 medium). After precipitating the tissue parts the medium was removed, 250 μl of trypsin/EDTA (0.25% trypsin, 1 mM EDTA-4Na; Gibco BRL, Life Technologies, Gaithersburg, MD, USA) was added and the tube was put into an incubator (37°C, 5% CO_2) for 45 min (Banker and Goslim 1998). The trypsin/EDTA was removed and 250 μl trypsin inhibitor/R12 [a concentration of 1 mg trypsin inhibitor (Gibco) per 1 ml R12 medium] and 50 μl DNase/R12 [a concentration of 1 μl DNAse (Deoxyribonuclease I, 151 U/μl; Gibco) per 250 μl R12 medium] were added. The tube was shaken for about 20 s until loose parts appeared whereafter 1 ml of R12 medium was added. After precipitating undissociated parts, the upper layer of the suspension was removed, put in a new tube, and centrifuged at a rate of 1,200 rpm for 5 min. The medium above the centrifuged cells was removed and the cells left at the bottom were mechanically dissociated (Romijn et al. 1984) using 1 ml of fresh R12 medium containing glutamine, vitamin C [0.025 g glutamine (Gibco, Scotland) plus 0.1 g vitamin C (Sigma-Aldrich Chemie Gmbh, Lemgo, Germany) dissolved in 10 ml sterile water was added to 1 l R12 medium] and NGF [10 μg NGF 2.5S (Sigma-) dissolved in 10 ml R12 medium was added to 1 l R12 medium]. All final neuron suspensions were counted using a Bürker chamber. Since the culture conditions do not favor the replication of nonneuronal cells, the cultures consist of more than 90% cortical neurons (Shahar et al. 1989).

The conductivity of the R12 medium used for all experiments, was measured to be nearly constant (1.6 S/m) for frequencies up to 10 kHz. Measurement of the viscosity of the medium using a "Contraves low shear" measurement system showed that the medium behaved like a Newtonic fluid ($\eta_{med} \approx 1*10^{-3}$ Pa s).

2.3
Theoretical Description of Dielectrophoretic Trapping

2.3.1
Estimation of the Dielectrophoretic Force

The two factors that largely determine the dielectrophoretic force are the electrical properties of cells and medium, and the gradient of the electric field. By using a single-shell model for the representation of the neuron and a finite element model for the calculation of the electric field, an estimation of the dielectrophoretic force created on a neuron positioned in the nonuniform electric field can be made.

The following values were assumed for the single-shell model of a cortical rat neuron (Pohl 1978; Kaler and Jones 1990): r=5 μm (measured); σ_{int}=0.75 S/m; $\varepsilon_{int}=\varepsilon_{med}=80*\varepsilon_0$ ($\varepsilon_0=8.85*10^{-12}$ F/m); $C_{mem}=18*10^{-3}$ F/m^2 ($C_{mem}=\varepsilon_{mem}/\delta$); σ_{med}=1.6 S/m (measured). The graph in Fig. 10 gives the real component of the Clausius–Mosotti factor as a function of frequency for different medium conductivities using Eqs. 2 and 3. Negative DEP will occur over the entire frequency range up to very high frequencies when the conductivity of the medium is 1 S/m or more as was already found (Fuhr et al. 1994; Markx and Pethig 1995). This has the advantage that the cells are distanced from damaging high field strengths, and a stable trapping position is created. Small changes in the values for conductivity and permittivity of the interior, the radius of the cell, or membrane capacitance did not result in significant changes of this curve (results not shown).

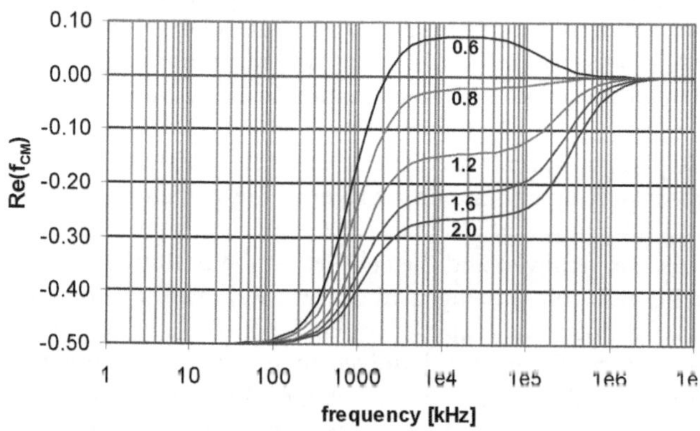

Fig. 10 Real part of the Clausius–Mosotti factor as a function of frequency for a neuron using a single-shell model. Each curve was calculated for a different medium conductivity: σ_{med}=0.6, 0.8, 1.2, 1.6, 2.0 S/m

Fig. 11 Column of medium meshed into tetrahedral elements consisting of 10 nodes of which the mid side nodes are not shown. *White lines* indicate the electrodes; the *circle* indicates the active region, which has a diameter of 380 μm

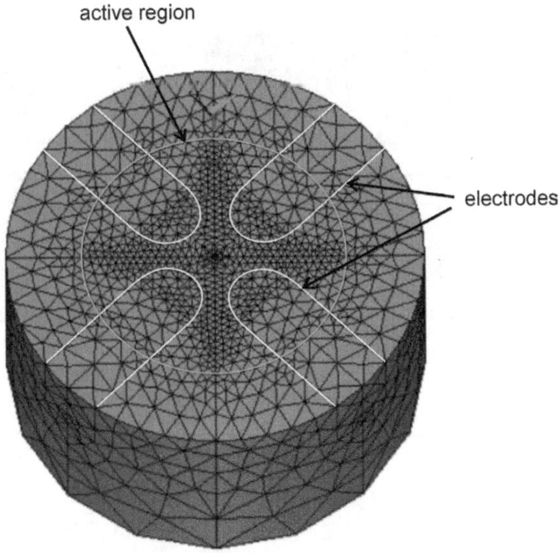

The gradient of the electric field is largely dependent on the geometry and the number of electrodes used. The finite element method (FEM) provides a formalism for generating discrete (finite) algorithms for approximating the solutions of differential equations. The complete electrode geometry can be modeled by subdividing (meshing) a finite-dimensioned subspace; in this case this subspace consisted of a cylindrical glass substrate on top of which a cylindrical volume of medium was present. Electrode-shaped areas that were shared by the substrate and medium elements represented the electrodes, since they were too thin (0.4 μm) to be modeled as volumes. Figure 11 shows the bottom view of the meshed medium with the electrodes indicated.

The frequencies used for the experiments ranged from 10 kHz to 50 MHz. These frequencies were low enough to result in a spatial distribution of the fields over the extent of the device comparable to that of static fields. Therefore, a quasi-static approximation was used, which means that only a subset of the Maxwell equations of electromagnetics needed to be solved.

ANSYS (ANSYS 5.6, Inc., Lake Forest, CA, USA) was used for the simulation of a finite element model. The finite element algorithm performs the computation at all nodes of the elements. From this solution the distribution of the field across the entire subspace is determined.

The result of the computation of the electric field is shown in Fig. 12 for 5 V_{pp} in the xy plane at z=0 μm for both electrode structures (Heida et al. 2002a). The dielectrophoretic force created by the field in the horizontal (z=5 μm) as well as the vertical plane (x=0) are shown in Fig. 12C and D, respectively. The maximum electric field strength was 85.1 kV/m and 141.8 kV/m in the horizontal plane at z=5 μm for an input signal of 3 and 5 V_{pp}, respectively. The maximum dielectrophoretic force created by these fields, assuming $|Re(f_{CM})|=0.5$, was 183.7 and 510.2 pN, respectively (r=5 μm, $\varepsilon_{r,med}=80$).

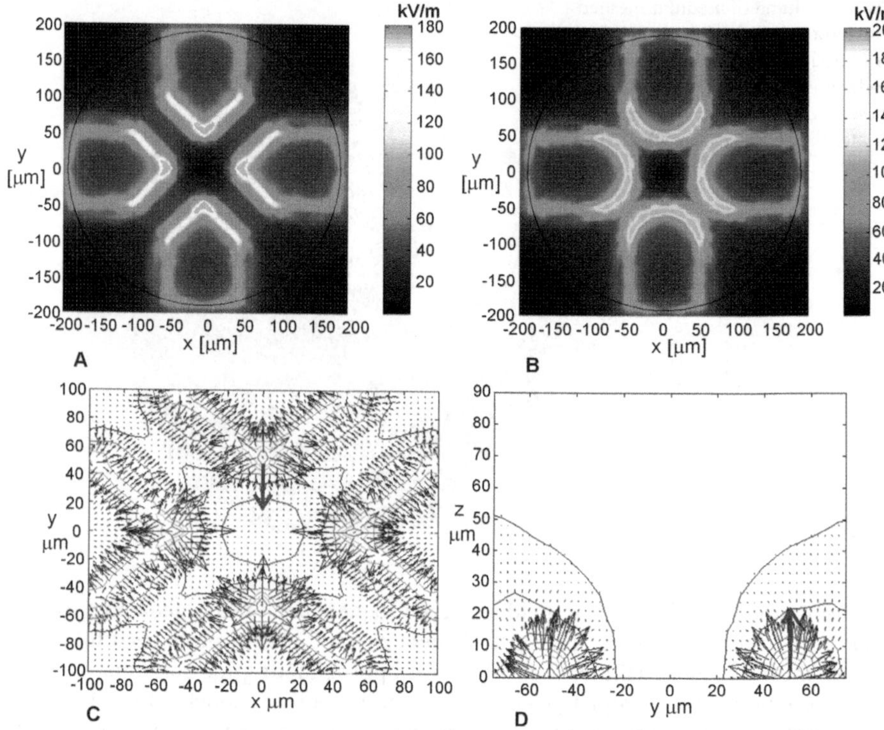

Fig. 12 The result of the calculation of the electric field by a finite element method in the xy plane at z=0 μm for an input signal of 5 V_{pp} for the electrode structure with **A** triangular shaped tips and **B** semicircular shaped tips. **C** The contours represent the equipotential field lines created by the quadrupole structure with triangular tips in the xy plane at z=5 μm, and **D** in the yz plane at x=0. The dielectrophoretic forces are indicated by *arrows*. The maximum dielectrophoretic force, indicated by a *bold arrow*, is 510 and 923 pN in the horizontal and vertical plane, respectively (Heida et al. 2002a)

2.3.2 Electrode–Electrolyte Interface

The arrangement of an electrolyte solution in connection with a metal electrode results in an unequal distribution of charge across the interface between both materials. Helmholtz suggested that two layers of electrical charge of opposite polarity are created; one layer of charge is located at the metal surface and the other, of equal but opposite charge, just inside the electrolyte. Between these two layers a potential difference exists. This "double layer" of charge at the interface behaves much like a parallel plate capacitor, termed the "double layer" capacitance, C_{DL}. However, some charge does manage to leak across the double layer due to electrochemical reactions taking place at the interface. Such charge leakage experiences a "charge transfer" resistance, R_{CT}.

When a direct current step is passed through a metal electrode, the voltage across the electrode–electrolyte interface increases until it reaches a value of the order of 1 V. At this level the electrode interface is no longer linear, and electrolysis will take

Table 1 Estimation of the actual field strength in relation to the theoretical field strength

Frequency [kHz]	$\frac{E_{estimation}}{E_{theoretical}} * 100\%$
10	44%
100	44%
1,000	88%
1,400	90%
4,000	91%
30,000	93%
50,000	93%

place (De Boer and Van Oosterom 1978). Therefore, AC signals are used for dielectrophoretic trapping experiments to avoid these unwanted electrochemical reactions.

At frequencies above 1 Hz it was observed that the interface impedance, Z_{el_med}, is well represented by a straight line at an angle ϕ to the real axis. This is given by the empirical equation (McAdams et al. 1995)

$$Z_{el\_med} = \frac{K}{(i\omega)^{\beta}} \tag{5}$$

K Measure of the magnitude of Z_{el_med}

β Constant which has a value between 0 and 1 (typically 0.8 for many biomedical electrode systems); $\beta=2\phi/\pi$

The electrode–electrolyte interface impedance of the electrodes of the quadrupole electrode structure was determined by impedance measurement. The characteristics of the results resembled the empirically found equation given above. Fitting of the measured electrode–liquid impedance resulted in K=1.5e8, β=0.9 (Heida et al. 2002a).

According to Eq. 5, at low frequencies the interface impedance is high and thus the electric field strength is decreased due to the potential loss over this impedance. With increasing frequency the impedance decreases and thus maximum electric field strength is achieved at these frequencies. Table 1 gives the percentage of the estimated field strength according to the prediction of the interface impedance for several frequencies in the range from 10 kHz to 50 MHz as used for the experiments.

2.3.3
Field-Induced Fluid Flow

When using a medium of high conductivity for DEP trapping, local Joule heating is inevitable. Joule heating, as computed in the electrical domain using an FEM model, was used as a volumetric source for the simulation of the same finite element model in the fluid domain to determine the temperature rise and the induced fluid flow.

All fluid properties were assumed to be isotropic. Density, specific heat capacity, viscosity, and thermal conductance were defined according to the properties of water (Lide 1994). The fluid was assumed to be incompressible, which means that the fluid density does not change with changing pressure p.

The following boundary conditions were applied in the fluid domain:

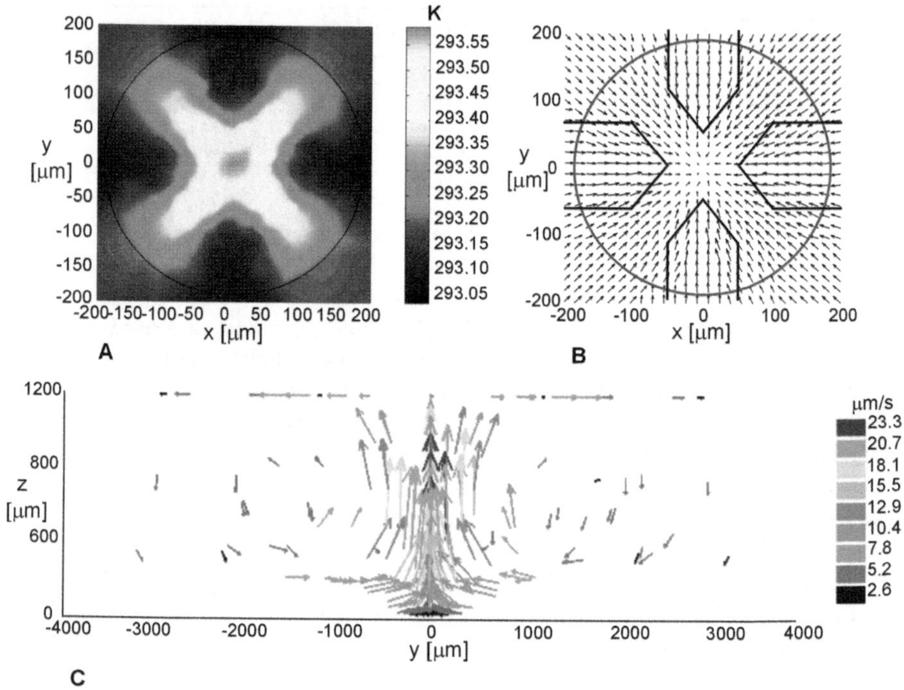

Fig. 13 **A** Local heating of the medium due to the electric field created by the quadrupole electrode structure with triangular shaped tips. The maximum temperature rise was induced in the center of the structure just above the substrate. **B** The fluid flow in the xy plane just above the substrate (z=5 μm). The maximum force is 0.007 pN in this plane. **C** The fluid flow in the yz-plane at x=0; the *length of the arrows* indicates the relative velocity (Heida et al. 2002a)

- Room temperature (293 K) at the vertical sides of the model and the bottom of the substrate
- Heat transfer coefficient of 50 at the upper surface of the medium (representing the convection between medium and air; Incropera and de Witt 1990)
- Zero velocity in z direction at the upper surface of the medium
- Zero velocity in x, y, and z directions at the medium/glass interfaces (no-slip "walls")

Figure 13 shows the temperature rise in the medium in the xy plane just above the substrate and in the yz plane at x=0 (Heida et al. 2002a). The fluid flow caused by this temperature rise is also shown in this figure in the horizontal plane at z=5 μm and in the vertical plane at x=0. The average velocity in horizontal direction at z=5 μm is 0.2 and 0.6 μm/s for 3 and 5 V_{pp}, respectively.

2.3.4
Total Trapping Force

It can be concluded that the dielectrophoretic force does not solely determine the trapping force. The total force on a cell consists of the summation of the dielectrophoretic force, the force due to the fluid flow and the gravitational force ($F_{tot}(x,y,z)=F_{DEP}+F_{flow}+F_g$). Of these forces the gravitational force is constant and was calculated to be 0.4 pN for a neuronal cell of 10 μm diameter.

The dielectrophoretic force is dependent on the electric field strength, which is dependent on the frequency in two ways. First, the real part of the Clausius–Mosotti factor is frequency dependent. Second, the electrode–electrolyte interface creates a frequency-dependent impedance. With increasing frequency $Re(f_{CM})$ decreases, while the interface impedance decreases resulting in an increasing electric field strength. The fluid flow is directly dependent on the electric field strength and therefore increases with increasing frequency.

2.4
Experimental Description of Dielectrophoretic Trapping

2.4.1
Experimental Procedure

The electrode plate was placed in a microscope setup (Nikon Inverted microscope, Diaphot-TMD, Nikon, Inc., Tokyo, Japan) and the connections at the two sides of the electrode plate were connected to a BNC connector and via a coaxial cable to the signal generator. A neuron suspension (20 μl, 10^6 cells/ml) was pipetted into a glass ring of 4 mm diameter and 1 mm height, which was positioned around the electrode structure while the electric field was applied. Three different amplitudes of the sinusoidal signal from the generator (1, 3, and 5 V_{pp}) in combination with ten different frequencies (10 and 100 kHz, 1, 4, 8, 12, 14, 18, 30, and 50 MHz) were used. By a digital video camera (Sony CCD Camera, Model DXC-151P, Sony, Inc., Tokyo, Japan) the experiments were time-lapse recorded. Every 10 s an image was taken during 30 min. Recording started directly after the application of the neuron suspension. For each setting of the amplitude and frequency three experiments were performed. Additionally, three experiments were performed in which no field was applied (reference experiments).

From the images, the number of cells trapped in the center, termed the "yield," was determined for all experiments. Neurons that did not stay in the center till the end of the experiment were not included in the yield. In addition, neuron density in the center as well as in the other areas inside the active region was qualitatively graded, as indicated in Figs. 8 and 9.

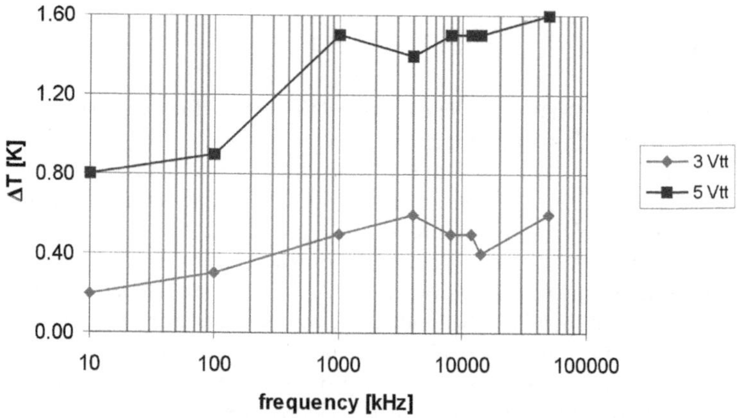

Fig. 14 Measured temperature rise as a function of frequency for 3 and 5 V_{pp}

2.4.2 Temperature Rise in the Medium Due to the Electric Field

Modeling results already showed that a fluid flow may be created as a result of Joule heating. Therefore, temperature measurements were performed just above the center of the electrode structure using the sensor of an ASEA Fiber thermometer 1010, an optical fiber with a diameter of 100 μm. Medium (20 μl at room temperature) was pipetted into the 4 μm diameter glass ring over which a drop of oil (Klearol–Witco B.V., Koog a/d. Zaan, The Netherlands) was applied to prevent evaporation.

Heating of the medium was found to be amplitude and frequency dependent. At 1 V_{pp} no temperature rise occurred at all. At 3 and 5 V_{pp} a temperature rise was noted during the first 5 min of field application. Thereafter it was stable irrespective of the duration of field application. Figure 14 shows the temperature rise after 5 min for different field settings (Heida et al. 2001a). At 5 V/10 kHz the temperature rise was 0.8°C. With increasing frequency the temperature rise increased. The maximum temperature rise was 1.6°C at 5 V_{pp}/50 MHz. At 3 V_{pp} the maximum temperature rise (at 50 MHz) was 0.6°C. It was found that the light from the microscope did not have any effect on the temperature.

2.4.3 Experimental Results

2.4.3.1 Trapping Neurons Under Various Field Conditions

Figure 15 shows the localization of trapped neurons after 30 min under four different field conditions. Fig. 15A gives the situation in which no field was applied. The other parts (B–D) show situations for three different amplitudes at 12 MHz.

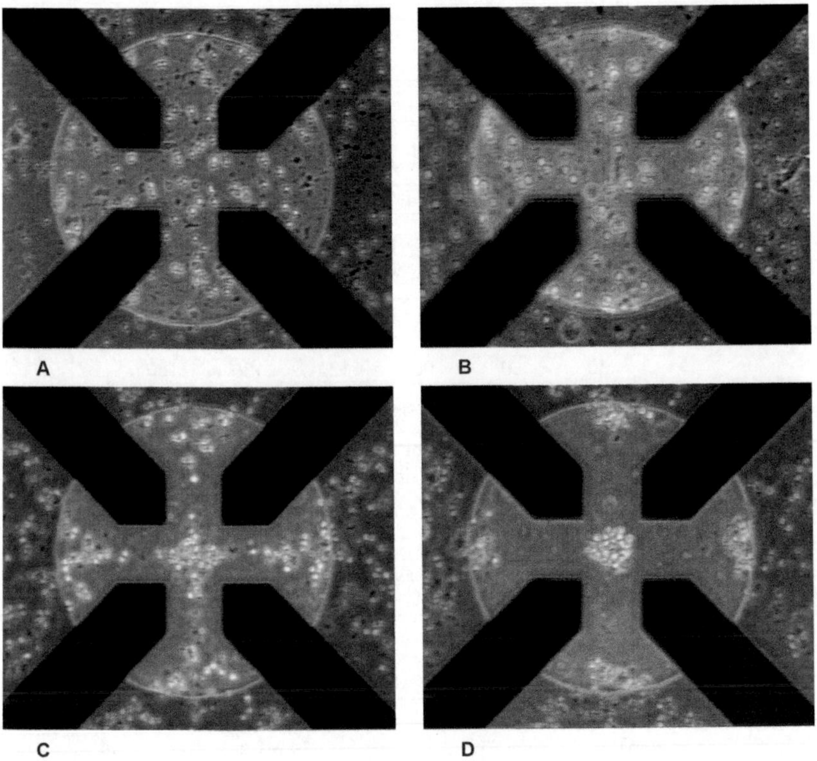

Fig. 15A–D Image number 180 of several experiments using different amplitudes (i.e., the situation after 30 min of field application). **A** No input signal (reference experiment), **B** 1 V_{pp}/12 MHz, **C** 3 V_{pp}/12 MHz, **D** 5 V_{pp}/12 MHz

In the absence of a field the neurons precipitated uniformly onto the surface. At 1 V_{pp} a uniform positioning of neurons was also obtained (image B). Neurons were trapped in the center of the electrode structure when the amplitude was increased to 3 V_{pp} (image C). In the lanes between the parallel sides of the electrode tips some neurons were also trapped at 3 V_{pp}. In the outer areas neurons collected in a triangular shape according to the shape of the diverging electrodes. Increasing the amplitude to 5 V_{pp} (image D) resulted in a distinct group of neurons collected in the center. The lanes were kept clear of neurons and the distance of neurons to the electrodes in the outer regions was clearly larger than at 3 V_{pp}.

2.4.3.2
The Yield

From the reference experiments it was learned that it took about 10 min for all neurons to precipitate. Figure 16 gives the average yield as a function of time at 14 MHz

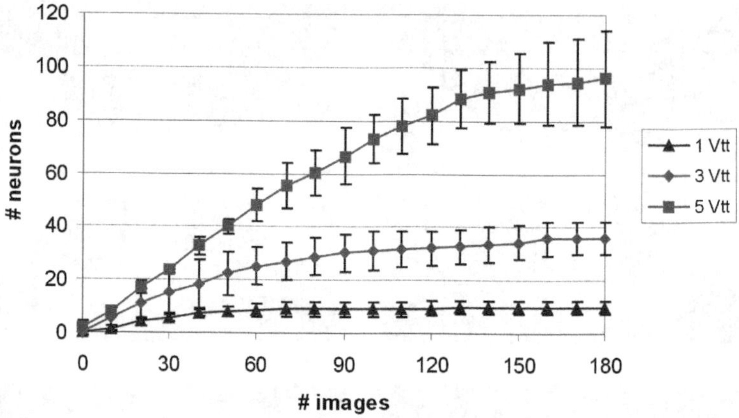

Fig. 16 Average yield of neurons as a function of time (10 images=100 s) at a frequency of 14 MHz for 1, 3, and 5 V_{pp}. The averages and standard deviations were calculated from the results of three experiments

Table 2 Minimum and maximum yield at 3 and 5 V_{pp}

Amplitude	Minimum yield		Maximum yield	
	Frequency	Neurons (n)	Frequency	Neurons (n)
3 V_{pp}	10 kHz	16	1 MHz	63
5 V_{pp}	4 MHz	36	14 MHz	114

for 1, 3, and 5 V_{pp}. Each data point represents the average accumulated number of neurons trapped in the center over 100 s (i.e., 10 images).

For almost all experiments the increase in yield was nearly zero during the last minutes of the experiment (most of the neurons were trapped during the first 20 min), indicating a saturation effect. The minimum and maximum yield for 3 and 5 V_{pp} is given in Table 2.

The variation in yield for each field setting was quite large. After 30 min of field application the average standard deviation was 26% for 5 V_{pp} and 16% for 3 V_{pp}.

Figure 17 shows the yield as a function of frequency (the DEP spectrum) after 20 min of field application. For all amplitudes a global peak was seen over the first part of the frequency range (10 kHz–4 MHz). The maximum was located at 1 MHz for 1 and 3 V_{pp}, and at 100 kHz for 5 V_{pp}. For frequencies in the range from 4 MHz to 50 MHz there was again a global peak in the 3 V_{pp}-curve. The maximum was located at 18 MHz. Over this frequency range several peaks were present in the 5 V_{pp}-curve.

In theory, without a field eight neurons would randomly be distributed in the center. The reference experiments showed an average of nine neurons in the center. The results of the 1 V_{pp} experiments were comparable to this zero-field "yield." Based on this random distribution and the assumption that the creation of the DEP funnel is restricted to the area between the electrode tips (assumed to be circular with a diam-

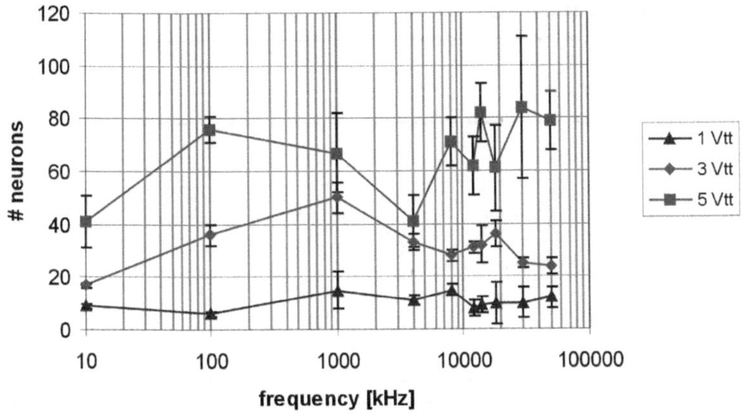

Fig. 17 The yield as a function of frequency after 20 min of field application

eter of 150 μm) about 28 cells would randomly be distributed over this area under zero-field condition. Thus, by using DEP, a maximum of 28 cells could theoretically be trapped in the center.

At 3 V_{pp} less than 28 cells were trapped in 30 min at 10 kHz and at 30 and 50 MHz. In this case, however, more than ten cells were detected in the lanes between the parallel sides of the triangular electrode tips, as indicated in Fig. 9. So, in total still about 28 cells were located within the 150-μm circular area. At 5 V_{pp} more than 28 cells were always trapped. This would not be possible if only gravitational and DEP forces were present. This enhancing effect was already stated in literature, and was attributed to the induced fluid flow (Müller et al. 1996).

The temperature rise is directly related to the electric field strength, which on its turn is directly related to the electrode–medium interface impedance. Since this impedance is largest at the lower frequencies the field strength increases with increasing frequency. In accord, the temperature rise was measured to be largest at 5 V_{pp}, and it increased with increasing frequency. The induced fluid flow is directly related to this temperature rise, and might therefore explain the enhanced trapping effect at 5 V_{pp}, which increases with increasing frequency as clearly shown in Fig. 17.

Strong variations in the yield occurred between the three experiments for each field setting. In earlier research large variations were also found. Pohl (1978) estimated that a scatter of some 15%–20% in the reproducibility of his yield existed. Differences in cell properties, cell diameter, and surface charge density can be ascribed to the large variation in yield (Huang et al. 1992; Green and Morgan 1997).

2.4.3.3
Qualitative Aspects of Neuron Trapping

As can be seen in Fig. 15C, the neurons could be packed quite closely together in the center. The possibility of neuron location in the lanes depended on the field parameters (compare image C and D of Fig. 15). The degree of compactness in the center

Table 3 Compactness of neuron clustering in the center and the neuron density in the lanes

Frequency	3 V_{tt}		5 V_{tt}	
	α	β	α	β
10 kHz	–	+	0	0
100 kHz	0	0	+	0
1 MHz	0	0	+	–
4 MHz	0	0	+	–
8 MHz	0	0	+	–
12 MHz	0	0	+	–
14 MHz	0	+	+	–
18 MHz	0	0	+	–
30 MHz	–	+	0	–
50 MHz	–	+	+	–

α Compactness of neuron clustering in the center, + compact clustering with possible piling up of neurons, o no piling up of neurons (confluent layer), – no confluent layer of neurons, β neuron density in the lanes, + >10 neurons located in the lanes, *0* <10 neurons located in the lanes, – no neurons located in the lanes

and the neuron density in the lanes were qualitatively graded as seen in Table 3 (the 1 V_{pp} experiments were left out).

Compact clustering of neurons in the center of the electrode structure went together with lanes that were kept clear of neurons. Moreover, compactness of the group of trapped neurons was amplitude-dependent: the higher the amplitude the closer the neurons were clustered. Increasing amplitude results in a stronger DEP force, and since it is directed upward and sideward near the electrodes, an increase in the force effectively results in a decrease of the "width" of the lanes. Neurons that formerly could precipitate in the lanes are now levitated. If next to the DEP force a fluid flow is present, levitated neurons will be swept away and (partially) directed to the center, enhancing the process of piling up in the center and clearing the lanes. Only for the 5 V_{pp} situation were the upward DEP force and the flow-related force strong enough to enhance the trapping effect as described in the previous section. However, only the fluid flow directed towards the center as shown in Fig. 13B can be responsible for the transportation of cells from outside the circular area of 150 μm diameter into the center.

2.4.3.4 Additional Considerations

It has to be kept in mind that the derivation of the expression for the DEP force (Eq. 1) was based on several assumptions as outlined in Sect. 2.1.1, i.e., Eq. 1 is the simplest approximation of the dielectrophoretic force. When the field is spatially highly nonuniform in comparison to the size of the cell or when the cell is located near a field null the assumption that the field is homogeneous in the vicinity of the sphere is no longer valid. Higher-order moments may be induced in the cell (Washizu 1992; Wang et al. 1994; Voldman et al. 2001).

However, it was found that when the particles are smaller than the electrode spacing, the dipole approximation gives an accurate result (Schnelle et al. 1999). Additionally, multipolar forces show similar frequency dependencies and are therefore hardly detectable. Errors of 1%–5% were found depending on particle size (Schnelle et al. 1999). The influence of higher-order terms was therefore negligible for the experimental situation described in this review.

Another aspect is mutual dielectrophoresis. The derivation of the dielectrophoretic force is based on a single cell. Nevertheless, cells disturb the field locally with the result that the dielectrophoretic force is altered such that attractive forces on other cells may be created. It has been found that mutual dielectrophoresis may be responsible for a rather enhanced trapping effect (Schnelle et al. 1996, 1999).

The ratio between inter-particle dipole–dipole force and the dielectrophoretic force acting on a single cell can be calculated according to (Dussaud et al. 2000; Jones 1986):

$$F_{ratio} \sim 6|f_{CM}|\psi \frac{\Delta}{d_{cell}} \qquad (6)$$

with ψ the volume fraction of the cells; Δ the characteristic length over which the field varies; $d_{cell} \sim r/\psi^{1/3}$ with r the radius of the cell. For the initial cell density (10^6 cells/ml) this ratio is about 1.4×10^{-7} indicating that the inter-particle interaction is negligible. When the cells precipitate and reach the substrate this cell density is raised locally, whereby F_{ratio} increases. However, since the interaction of the closely spaced particles is due to their field-induced polarization, the interaction force decreases with decreasing field strength. In the center of the electrode structure the field is minimal, as can be seen in Fig. 12. The inter-particle interaction is therefore negligible here. In addition, adhesive forces have values in the nN range and would therefore dominate possible field-mediated polarization forces, since these have values in the lower pN range (Glasser and Fuhr 1998).

Lastly, it is possible to discriminate between cells on the basis of their physiological state. For example, DEP spectra of viable and nonviable yeast cells show substantial differences (Huang et al. 1992). Furthermore, differences between normal and cancerous cells (Gascoyne et al. 1994) have been characterized. Since the maximum temperature rise was only 1.6°C for the higher frequencies this change was not expected to have harmful effects on the viability of the cells. However, using frequencies below 1 MHz may cause irreversible damage due to loading of the membrane. Therefore, the single-shell model as used for the whole frequency range may only be applicable to the upper part of this range.

3 Exposing Neuronal Cells to Electric Fields

The creation of dielectrophoretic forces requires electric fields of high nonuniformity. High electric field strengths, however, may cause excessive membrane potentials by which cells may be changed irreversibly or lead to cell death.

Membrane breakdown was investigated theoretically as well as by experiment, for frequencies up to 1 MHz.

3.1 Theory

3.1.1 Membrane Breakdown

Membranes play a crucial role in living cells and organisms. The outer membrane of biological cells separates the exterior, nonliving environment from the internal, living, cytoplasm. It provides a selective diffusion barrier to molecules moving into and out of the cell. Additionally, it contains energy driven transport mechanisms that maintain large concentration differences of many substances between the external environment and the cytoplasm. For example, in the cytoplasm of most animal cells the concentration of Na^+ is much less and the concentration of K^+ much greater than the extracellular concentrations of these ions. This situation is brought about by the action of a Na^+, K^+ pump in the membrane that pumps Na^+ out of the cell and K^+ into the cell. This pump activity is the result of an integral membrane protein, the Na^+, K^+-ATPase.

Apart from the outer membrane surrounding the cell, many of the internal organelles are enveloped by membranes. Almost all of the biochemical processes in the living cell are associated with membrane-bound enzyme and transport mechanisms.

The cell membrane consists of a bimolecular layer of lipid molecules, the lipid bilayer. This bilayer provides a fluid matrix in which membrane proteins are embedded. The electrical properties of cell membranes are mainly determined by these proteins (Zimmermann and Neil 1996).

The large concentration differences between exterior and interior determine the resting membrane potential. The maintenance of a set membrane potential is of importance in controlling the amplitude of nerve impulses (action potentials) and muscle contractions, while the energy driven pumps or carriers play a supporting role in electrical signaling.

Under a DC field, application an "additional" membrane potential is induced. When assuming that this field is homogeneous, that the resting membrane potential is superimposed on the field-induced membrane potential which is not changed by the external field, and that surface admittance and space charge effects do not play a role, the membrane potential can be calculated according to (Zimmermann and Neil 1996, Kotnik et al. 1997):

$$V_m = f_s E r \cos\alpha (1 - \exp(-t/\tau)) \pm V_{rest} \tag{7}$$

with E the static electric field, V_{rest} the resting membrane potential, r the cell radius, α the angle between the field line and a normal from the center of the sphere to a point of interest on the membrane (the generated potential is maximal at membrane sites oriented in the field direction α=0° or 180°, and zero at sites perpendicular to the field), t is time, τ is the time constant of the membrane, and f_s is a function reflecting the electrical and dimensional properties of the cell and the surrounding medium; f_s can be calculated according to:

$$f_s = \frac{3\sigma_{med}\left[3\delta r^2 \sigma_{\text{int}} + \left(3\delta^2 r - \delta^3\right)(\sigma_{mem} - \sigma_{\text{int}})\right]}{2r^3(\sigma_{mem} + 2\sigma_{med})(\sigma_{mem} + 0.5\sigma_{\text{int}}) - 2(r-\delta)^3(\sigma_{med} - \sigma_{mem})(\sigma_{\text{int}} - \sigma_{mem})} \tag{8}$$

where σ_{med}, σ_{int}, and σ_{mem} are the electric conductivities of the medium, of the cytoplasm, and of the membrane, respectively, and δ is the membrane thickness.

The time constant of the membrane can be written as:

$$\frac{1}{\tau} = \frac{1}{R_{mem} C_{mem}} + \frac{1}{r C_{mem}\left(\frac{1}{\sigma_{\text{int}}} + \frac{1}{2\sigma_{med}}\right)} \tag{9}$$

with R_{mem} and C_{mem} the specific resistance and capacitance of the membrane per unit area, respectively; r is the cell radius, and σ_{int} and σ_{med} are the specific conductivity of the cell interior and medium, respectively.

If the resting membrane potential is sufficiently high, asymmetric breakdown of only one hemisphere may be expected. Assuming that V_m is directed towards the interior of the negatively charged cell, the vector of the field generated across the membrane is parallel to the membrane field vector in the hemisphere facing the anode, but antiparallel to the hemisphere facing the cathode. Thus, membrane breakdown in the anodic hemisphere requires lower external field strengths than in the membrane facing the cathode, termed asymmetric breakdown.

Equation 7 can be simplified under the following assumptions:

- The time constant of the membrane, τ, is much smaller than the duration of field exposure, i.e., pulse duration $[(1-\exp(-t/\tau) \rightarrow 1]$
- The membrane conductivity σ_{mem} can be neglected (f_s=3/2)
- The membrane thickness δ is much smaller than the cell radius r
- When the induced membrane potential is sufficiently high the resting membrane potential can be neglected

The expression then becomes (Marszalek et al. 1990, Zimmermann and Neil 1996):

$$V_m = 1.5Er\cos\alpha \tag{10}$$

Figure 18 shows three theoretical situations in which the occurrence of asymmetric and symmetric breakdown is indicated.

When an alternating (sinusoidal) electric field with amplitude E is applied, the generated membrane potential is given by the following equation (Schwan 1957, Marszalek et al. 1990, Zimmermann and Neil 1996):

$$V_m = \frac{1.5Er\cos\alpha}{\sqrt{1+(2\pi f\tau)^2}} \tag{11}$$

with f the frequency of the field. Taking into account the assumptions made above, the time constant τ of the membrane can now be written as:

$$\tau = rC_{mem}\left(\frac{1}{\sigma_{\text{int}}}+\frac{1}{2\sigma_{med}}\right) \tag{12}$$

Equations 10 and 11 indicate that the induced membrane potential increases with increasing cell radius. It is therefore assumed that the cell organelles are unaffected by the breakdown of the cell membrane. The single-shell model is therefore valid for the investigation of membrane breakdown of biological cells.

The theory presented above is applicable only for small external electric fields. When large membrane potentials are induced the theory fails (DeBruin et al. 1999). The charging transient is interrupted and the membrane potential becomes nearly constant at about 0.2–1 V (Zimmermann and Neil 1996; Teissié 1993; Benz 1980a, b, c), most apparent at the poles of the cell, and the cosinusoidal profile of the induced potential around the cell is lost. At this point membrane breakdown or electropermeabilization occurs resulting in an increased membrane conductivity. Figure 19 shows an example for a cell exposed to a uniform electric field. The situation for electropermeabilization as well as the situation described by Eqs. 7 and 10 is presented.

Under nonuniform field conditions the calculation of the breakdown voltage using the equations presented above become rather unpredictable due to the dependence of the electric field on the x, y, and z coordinates. Accordingly, the value of the induced membrane potential becomes largely dependent on the position of the cell in the three-dimensional space.

3.1.2
Pulse Length, Temperature, and Medium Conductivity Dependence

The high permeability state caused by dielectric breakdown is dependent on the amplitude and duration of the electric field as well as on temperature and medium conductivity (Ryttsén et al. 2000; Benz et al. 1980a, b, c; Zimmermann and Neil 1996).

For example, it was found for lipid bilayers and cell membranes that at a pulse length of 10 μs the breakdown potential of was of the order of 0.5 V; it rose to approximately 1.2 V for a pulse length of 800 ns. For pulse lengths longer than 10 μs or

Fig. 18 Schematic diagram of the induction of symmetric (first and second) and asymmetric (last) membrane breakdown in a spherical cell of radius a, exposed to a stationary external field; α is the angle between the membrane patch and the external field. The breakdown level is first reached in field direction, α=0° or 180°, i.e., $E=E_b$. The resting membrane potential is assumed to be negligible in this case. With increasing field strength, $E>E_b$, breakdown also occurs at the membrane sites oriented at a certain angle, α, to the field direction and pores are created. If the resting membrane potential, V_{rest}, is taken into account an asymmetric breakdown is expected. For the case in which the cell interior is negatively charged with respect to the external medium, the generated membrane potential, V_m, is parallel to the membrane potential

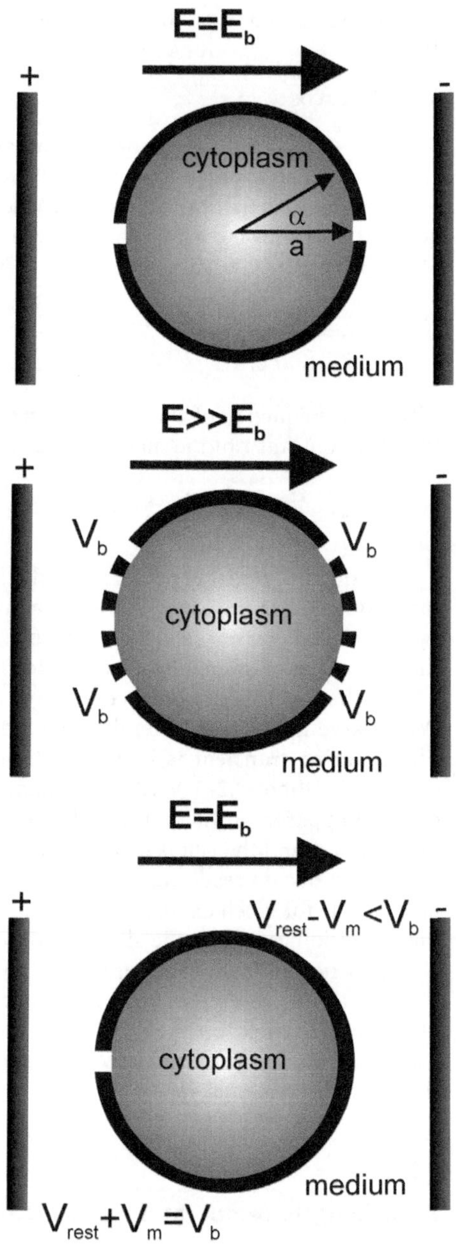

shorter than 800 ns the breakdown potential remained constant at 0.5 and 1.2 V, respectively. These values applied to the situation in which the temperature was about 25°C. A pulse length of several hundred ns at 4°C resulted in a breakdown potential of 2 V, while at 40°C it was only 0.5 V (Zimmermann and Neil 1996, Benz et al. 1980a).

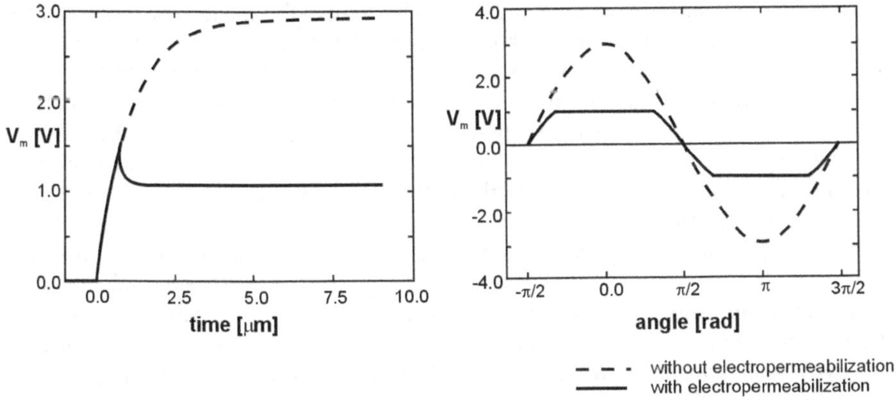

Fig. 19A, B Example of the membrane potential V_m of a cell exposed to a uniform electric field. **A** The membrane potential rises until breakdown occurs, after which the membrane potential is kept constant. **B** The cosinusoidal profile of the membrane potential as a function of the circumference of the spherical cell. The membrane potential cannot attain a higher value than breakdown level and the cosinusoidal profile is lost

These results indicate that the breakdown phenomenon occurs in less than 10 ns. The specific conductivity of the cell membrane can rise from about 10^{-3} S/m^2 to more than 10×10^3 S/m^2 (Coster 1999; Benz et al. 1980c).

In addition, the degree of membrane permeabilization at breakdown was found to be critically dependent on the conductivity of the medium; low-conductivity conditions result in an increased permeability (Ho and Mittal 1996; Sukhorukov et al. 1998).

3.1.3
Pore Model

Lipid bilayers have a central, nonpolar, layer, which is about 2–3 nm thick. Such a thin sheet of low permittivity (ε~2) excludes ions and charged molecules. The electrostatic energy needed to move charge from a high dielectric medium (such as water, ε_{water}=80) into a low dielectric medium, termed the Born energy, is mainly due to the large difference in permittivities of both materials (Coster 1999). The energy needed to insert a small ion into a membrane is significantly reduced if the ion is placed into a (mobile) aqueous cavity or if it can pass through an aqueous channel, i.e., aqueous pathways (Weaver 2000). The greater reduction is achieved by the pore and is therefore assumed to form the basis of electroporation/electropermeabilization.

The pores will contribute to the electrical conductance of the membrane. In fact, the conductance of the bilayer itself is so small that the conductance is entirely due to that of the pores, and thus is greatly determined by the number of pores. The conductance of a pore filled with medium will depend on the radius of the pore as well as the concentration of ions.

Pores allow transport of ions and molecules across the membrane, however, the transport mechanism is not fully understood yet (Benz 1980c, Weaver 2000). Electrophoresis, diffusion, and endocytosis are possible mechanisms (Weaver 1993).

A heterogeneous distribution of pore sizes is expected due to thermal fluctuations and variations in electric field energy within the membrane (Weaver 2000). Once the energy required to form a pore is known, the relative probability of formation of pores of different radii, according to the Boltzmann distribution function, can be determined (Coster 1999).

Pores experience a progressively smaller expanding force as they expand. Just outside a pore's entrance and exit there are inhomogeneous electric fields and an associated "spreading resistance," such that as the pore grows, a progressively greater fraction of the membrane potential appears across this spreading resistance. Accordingly, less voltage appears across the pore itself and, therefore, the electrical expanding pressure is less. The minimum pore size (R_{min}) is about 1 nm; thus, the small ions that comprise physiologic saline can be transported through the pores. Additionally, a critical radius (R_c) exists, meaning that a pore of that size will spontaneously grow and rupture the membrane.

It is assumed that hydrophobic pores are formed first due to the reorientation of the dielectric dipoles of the lipid molecules. These pores transform to hydrophilic pores, termed the primary pores, with larger radius because the energy cost to make the pore circumference ("edge energy") is much larger for hydrophilic than for hydrophobic pores (Weaver 1993). Hydrophilic pores are believed to be metastable for short periods of time (Tsong 1991; Weaver 1993). Very large pores, secondary pores, may evolve from the primary pores because of pressure driven flow(Weaver 1993).

It was thought that larger membrane potentials, such as those created near the poles according to Eqs. 9 and 10, produce more pores to shunt the extra stimulus current across the membrane (DeBruin 1999). Increasing the electric field strength results in a larger fraction of the cell membrane being electroporated. However, only a small fraction of the membrane surface is generally affected. For example, the maximum fractional area of the membrane occupied by the pores was found to be about 10^{-4}–10^{-3} (DeBruin 1999).

The aqueous pore theory enables us to describe the transient response of the cell membrane to an electrical pulse. It consists of the following chronological phases (Tsong 1991; Chang 1990; Ryttsén 2000):

- The increase in membrane conductance appears immediately or a few milliseconds after the pulse.
- Expansion/evolution of the pores appears during the first few milliseconds at a rate of 3.1–7.3 pA/ms.
- The high-conductance state of the membrane lasts up to about 100 ms.
- The recovery process consists of two sub-phases:
  (a) A fast recovery process (in the millisecond range) in which the conductance returns to the approximate prepulse level, pores shrink and partially reseal.
  (b) A slow recovery process (several seconds to minutes) during which the low-conductance state turns into a condition of pore closure.

3.1.4
Electromechanical Model

In spite of the ability of the aqueous pore theory to explain the electropermeability state and its transience, this theory does not include electromechanical loading. Assuming that the membrane can be regarded as a capacitor filled with a homogeneous elastic dielectric material, it can be postulated that the thickness of the membrane capacitor depends on the electrical compressive and mechanical forces (e.g., internal hydrostatic pressure; Zimmermann and Neil 1996). For example, in high-intensity electric fields, dipoles arise from the free charges that accumulate at the two cell interfaces, i.e., at the inner and outer boundaries of the membrane. These are Maxwell–Wagner interfacial polarizations and the stresses these cause result in deformation forces that are proportional to the real part of the Clausius–Mosotti factor (Sukhorukov et al. 1998). Such deformations may lead to changes in the membrane structure without the formation of distinct pores, a possible explanation of the high permeability state of the membrane at breakdown level. Electromechanical compression of the membrane material, when sufficiently large, could, for instance, expose transmembrane conduction molecules that are normally buried in the membrane matrix. This would result in a sharply increased electric conductance through the membrane.

The electromechanical model is able to explain some of the observations that are difficult to interpret by the aqueous pore theory (Zimmermann and Neil 1996; Sukhorukov et al. 1998):

- The dependence of the dipole generation on medium conductivity accounts for the observed dependence of electropermeabilization on medium conductivity.
- By assuming that the membrane material is not perfectly elastic and shows inertia, the pulse-length dependence of the breakdown potential can be explained. If the elastic compressive modulus transverse to the membrane plane is a function of the compression time, it increases with increasing rate of compression, comparable to other visco-elastic materials.

3.1.5
Recovery of the Membrane

The recovery process of the membrane was found to be temperature-dependent. For example, at 4°C complete recovery may require about 30 min; at 20°–30°C it may require significantly less time (Zimmermann and Neil 1996). These temperature effects reflect the fact that active membrane processes require optimal temperatures for the function of enzymes (Zimmermann and Neil 1996).

The recovery of the membrane to its original low conductivity and low permeability state does not necessarily mean that the original membrane structure is completely restored (Zimmermann and Neil 1996). The concentration and nature of the pulse medium ingredients play an important role in membrane recovery. In general, pulse media should consist of ultrapure substances dissolved in double distilled water.

Cell death may occur due to rupture of the membrane or chemical imbalances resulting from influx and efflux during the high permeability state of the membrane (Weaver 1993).

3.1.6 Methods of Observation

The high permeability state of the membrane can most effectively be measured by staining the cell membrane with appropriate fluorescent-labeled dyes (Ho 1996; Zimmermann and Neil 1996; Tsong 1991). Attempts have been made to visualize pores by electron microscopy (Chang and Reese 1990). However, visualization of (primary) pores by any present form of microscopy seems to be unlikely, because of their small size and short lifetime (Weaver 1993, 2000; Zimmermann and Neil 1996). Fixation or freezing of freely suspended cells that were exposed to electric field pulses, requires milliseconds to seconds, which is rather long. Additionally, it is very likely that the "pores" seen in electron microscopic images of cells electropermeabilized in iso-osmolar solutions are artifacts introduced during the freezing process by the loss of the osmotic barrier.

To test "normal'" functioning of the cell after field exposure, parameters such as cell growth, division rates, viability, and cell motility should be used as these integrate influences over long observation times.

3.2 Theoretical Investigation of Induced Membrane Potentials of Neuronal Cells

3.2.1 The Model

As was shown in Sect. 2, finite element modeling can be used to determine the electric field inside a three-dimensional medium. Into the model described in Sect. 2, a neuronal cell was introduced by modeling it as a sphere with homogeneous properties. The surface area of this sphere, which was shared by the surrounding medium and the sphere, was meshed. This meshed area was copied and used for the definition of a new sphere. This second sphere represented the cytoplasm. The first sphere was deleted while keeping the area mesh on the medium intact. Two layers of nodes, with exactly the same positions without being connected, were the result. One layer belonged to the medium and one layer belonged to the cell interior (Fig. 2). Circuit elements were introduced between two nodes located at the same position. These elements represented the membrane, because the membrane is too thin to be modeled as a volume. Electrically, the membrane consists of a resistance in parallel with a capacitance. Each node covered a small part of the spherical surface and this area determined the local resistivity and capacity of the neuronal membrane (R_{mem}=0.1 Ωm^2, C_{mem}=9 mF/m^2). For the computation of the electric field and the

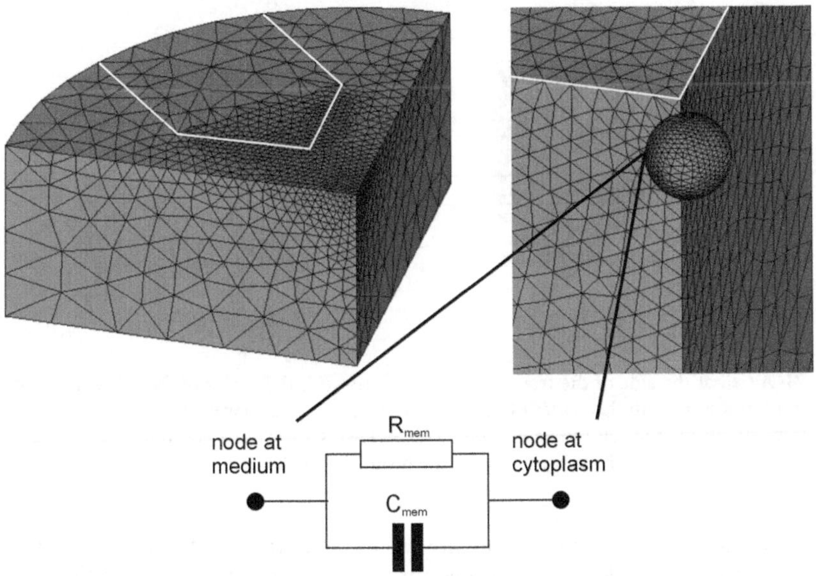

Fig. 20 "Element plot" of the neuronal cell and the surrounding medium (bottom view). A sub model consisting of a quarter of the complete electrode structure was used for accurate computation of the electric field around the cell and the corresponding induced membrane potential. *White lines* indicate the contour of the electrode tip. On the *right*, a part of the electrode tip with the cell is presented. The *lower* figure schematically shows the parallel combination of a resistance, R_{mem}, and capacitance, C_{mem}, representing the membrane, which created the connection between two nodes located at the same position, i.e., the node in the medium and the node in the cytoplasm. For computation of the membrane potential, the magnitude of the parallel combination was used incorporating the field frequency, which was modeled as a resistive element

membrane potential the magnitude of the resulting impedance was calculated for each frequency and represented by a resistive element.

The complete model included all four electrodes. Therefore, after simulation of the complete model a second simulation was performed using a sub model, which consisted of a quarter of the model. Element sizes could be scaled down with this sub model so that a high accuracy was attained in the computation of the electric field around the cell and the induced membrane potential. Figure 20 shows the elements of the sub model.

Again, areas represented the electrodes. Applying opposing but equal potentials, e.g., –2.5 and +2.5 V, per two neighboring electrodes the potential distribution and the electric field were computed. The potential difference over the parallel combination of membrane resistance and capacitance represented the membrane potential.

Since the membrane area affected by membrane breakdown is strongly dependent on local electric field intensities, a cell located at the tip or along the side of the electrode was expected to show different results in breakdown experiments. Two models were used each including one cell; Figure 21 shows the locations of the two cells.

It has to be kept in mind that both Eq. 11 used for the calculation of V_m, and the finite element model are based on the assumption that the membrane capacitance is

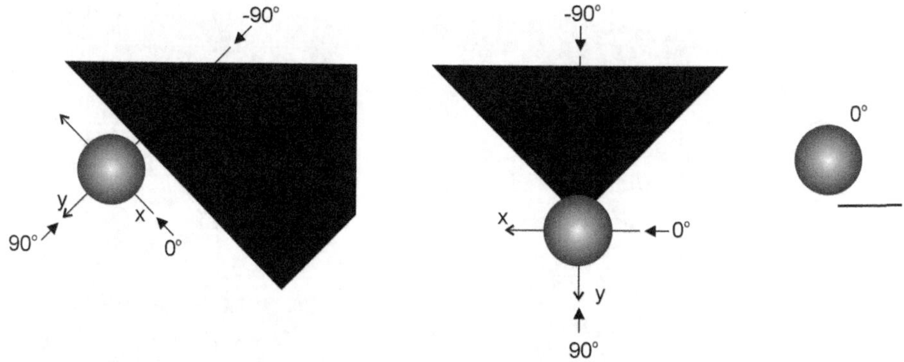

Fig. 21 **A** Cell at the side of the triangular electrode tip with the center of the cell at z=7 μm. **B** Cell at the electrode tip with the center of the cell at z=8 μm. **C** Side view of the cell above the electrode. For both situations the cell radius is 6 μm. Three viewing directions are indicated as −90°, 0°, and 90°. These were used for the presentation of the results in Fig. 23

constant. In reality, the specific membrane capacitance is determined by the dielectric constant and thickness of the membrane. Almost independent of the density of proteins in the membrane, the specific capacitance of neuronal membranes was found to be 9 mF/m^2 (Gentet et al. 2000). Nevertheless, the relaxation frequency for proteins and lipids, the main constituents of the membrane, is in the range from 100 kHz to 10 MHz (Pohl 1978). In practice, frequency dependency could therefore also be the result of additional dispersions in the membrane. However, protein dispersions are masked by the stronger β or Maxwell–Wagner dispersion (Gimsa and Wachner 1998). If a constant membrane permittivity and conductivity is assumed, the models used are justified.

3.2.2
Modeling Results

With a cell located near the electrode the electric field is locally disturbed. At low frequencies the membrane completely blocks the electric field, while at frequencies above 1 MHz the membrane potential drops and the field penetrates the cell interior. A higher field strength induces a larger membrane potential. The distribution of membrane potentials is shown in Fig. 22 for the two cells (Heida et al. 2002b).

According to these simulations it was found in both situations that a large membrane potential is induced at only a small part of the membrane. Due to the large field gradient, the variation in membrane potential is largest for the cell at the tip. Figure 23 shows the area of the membrane for which the membrane potential exceeds a certain threshold (the total membrane surface for a cell with a radius of 6 μm is 452×10^{-12} m^2). The threshold ranges from 0.2 to 1 V.

While the maximum membrane potential of a cell located along the side of the electrode was 0.75 V compared to a maximum of 1 V for the cell at the tip, the total area affected by the field was equal in size for cells located at either location for low threshold values (0.2–0.4 V).

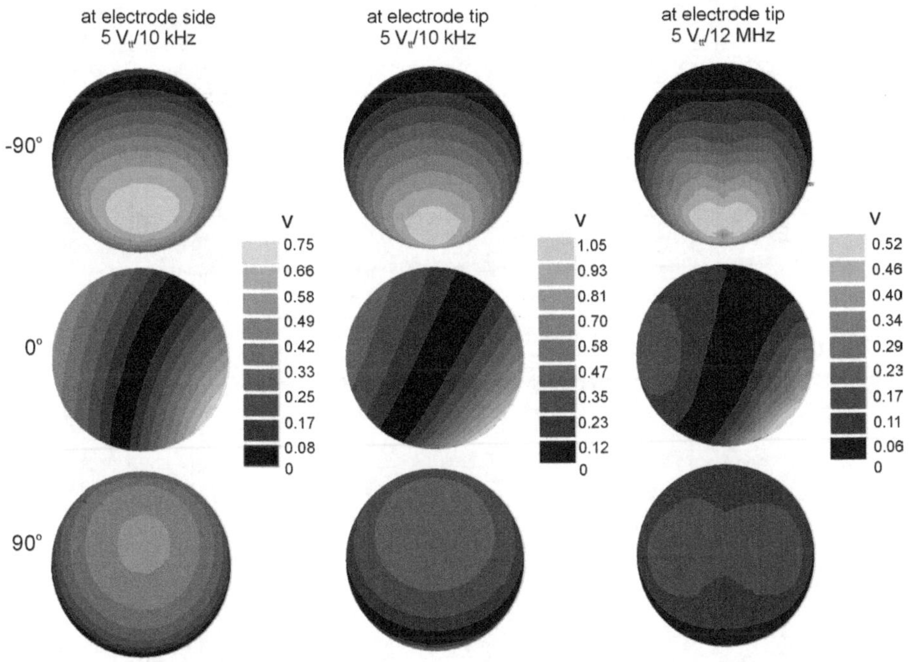

Fig. 22 Membrane potentials determined for a cell positioned along the electrode side at 5 V_{pp}/ 10 kHz (**A**), and a cell positioned at the electrode tip at 5 V_{pp}/10 kHz (**B**), and 5 V_{pp}/12 MHz (**C**) (Heida et al. 2002b)

3.3 Experimental Investigation of Neuronal Membrane Breakdown

3.3.1 Experimental Procedure

A drop of cell suspension (1.5×10^6 viable cells/ml) was applied on top of an electrode structure. In order to investigate membrane breakdown high field strengths are needed in the vicinity of the cell. Application of an electric field directs the cells to regions of minimal field strength by negative dielectrophoresis. Therefore, to allow the cells to precipitate randomly and adhere to the substrate, the plate was set at rest for 30 min. Due to the adhesive forces, the cells initially positioned close to the electrodes were thus not immediately pushed away from the electrode at field application.

After these 30 min the electrode plate was clamped into a small box in which small pins provide the connection from the gold electrodes to the coaxial cable leading to the signal generator. This box was directly placed on a Nikon Inverted microscope (Diaphot-TMD). An image was taken from the cells by a digital camera (Sony CCD Camera, Model DXC-151P). The electric field was then applied using a sinusoidal signal from a HP 3245 A universal source. Amplitude and frequency were set via a computer according to a Labview script (LabVIEW) in the range from 2 to 18 V_{pp}, and

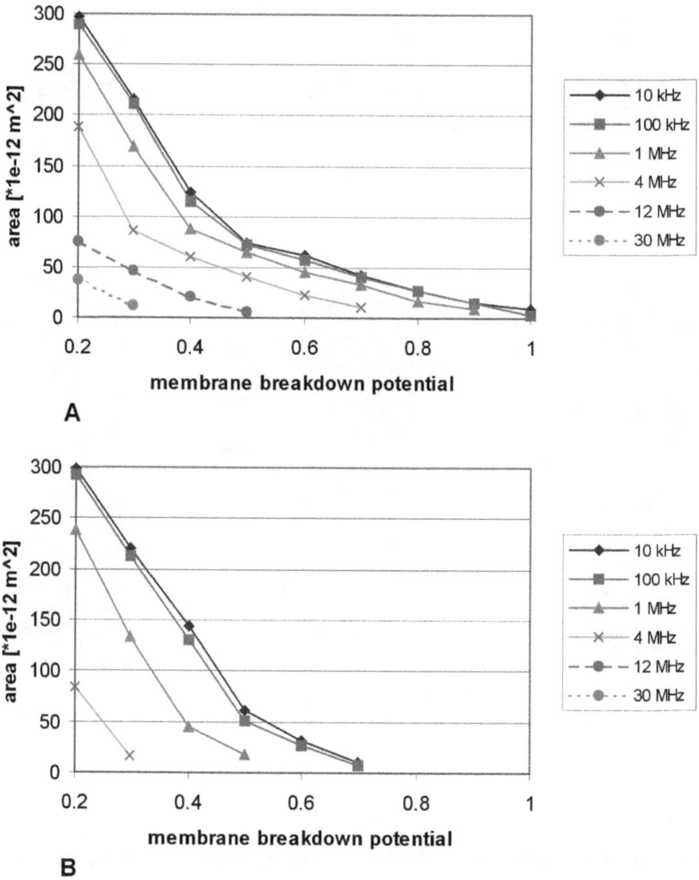

Fig. 23 Membrane area for which the membrane potential exceeds a certain value for a cell located at the tip (**A**) or along the side (**B**) of an electrode. The membrane potential ranges from 0.2 to 1 V

100 kHz to 1 MHz, respectively. The frequency was constant while the amplitude of the signal was increased in steps of 0.5 V_{pp} from 2 to 18 V_{pp}. Each amplitude was applied for about 14 s, i.e., 10 s and the time needed to take an image, which was about 4 s. Frequencies in the range from 100 kHz to 1 MHz were used.

Membrane rupture or cell collapse was detected using the principle of phase contrast microscopy. The moment a neuronal cell collapses it turns dark. This effect is comparable to the uptake of a staining substance like Trypan Blue. Experiments were performed in which some of the electrode structures were covered with a normal cell suspension, while others were covered with a drop of cell suspension to which a small drop of Trypan Blue (about 4 μl) was applied. These experiments confirmed that the use of phase contrast microscopy made the process of cell collapse visible with equal precision as compared to detecting the uptake of Trypan Blue.

Table 4 Estimation of the actual field strength in relation to the theoretical field strength

Frequency [kHz]	$\frac{E_{estimation}}{E_{theoretical}} * 100\%$
100	44%
250	70%
800	87%
1,000	88%

3.3.2 Data Analysis Procedure

The first and last images taken during the experiment were used to identify ten cells that collapsed while remaining at the same position during the experiment. Assuming the cells to be spherical, the diameter was determined from the first image and was assumed to remain constant until collapse. For each of the identified cells, the point at which collapse occurred was detected by scanning the images in sequence. Each image is related to a certain amplitude of the input signal and thus to a certain electric field strength near the cell. From the first image a point on the cell was indicated as being the membrane site at which the induced membrane potential was largest. The field intensity at this point was determined by finite element modeling, as described in the Sect. 2 and previous sections, from which the membrane potential was calculated according to Eq. 11, assuming cos α=1 (the maximum induced membrane potential).

In practice, the calculated maximum field strengths were not realized due to non-ideal material properties. Theoretical investigation, as described in Sect. 2, showed that even at high frequencies the maximum field strength induced in the medium was about 90% of the ideal value. Table 4 gives the maximum field intensities per frequency up to 1 MHz as a percentage of the theoretical maximum field strength. According to these estimations the maximum induced membrane potentials $V_{m,max}$ were calculated.

It has to be noted that at the point of collapse the membrane probably had already adopted a high permeability state, and therefore the calculated potential will tend to be somewhat higher than the factual breakdown level.

For the graphic presentation of the results, the percentage of cells that had collapsed as a function of the membrane potential was determined. The potential was taken from 0 to 2 V with intervals of 0.1 V. At each interval and for each frequency the number of collapsed cells was checked.

3.3.3 Experimental Results

Figure 24 shows four images of the experiment performed at 800 kHz. In general, cells located near an electrode collapsed at lower field intensities in comparison to cells located further away. The cells had an average diameter of 10 μm (SD=2).

Figure 25 shows the percentage of cells that have collapsed as a function of the membrane potential for all experiments. Most of the cells collapsed at a membrane potential of 0.4 V at 100 kHz. For higher frequencies the percentage of dead cells was smeared out over a range from 0.5 to about 1.1 for 250 and 800 kHz, and from 0.6 to 1.7 V for 1 MHz.

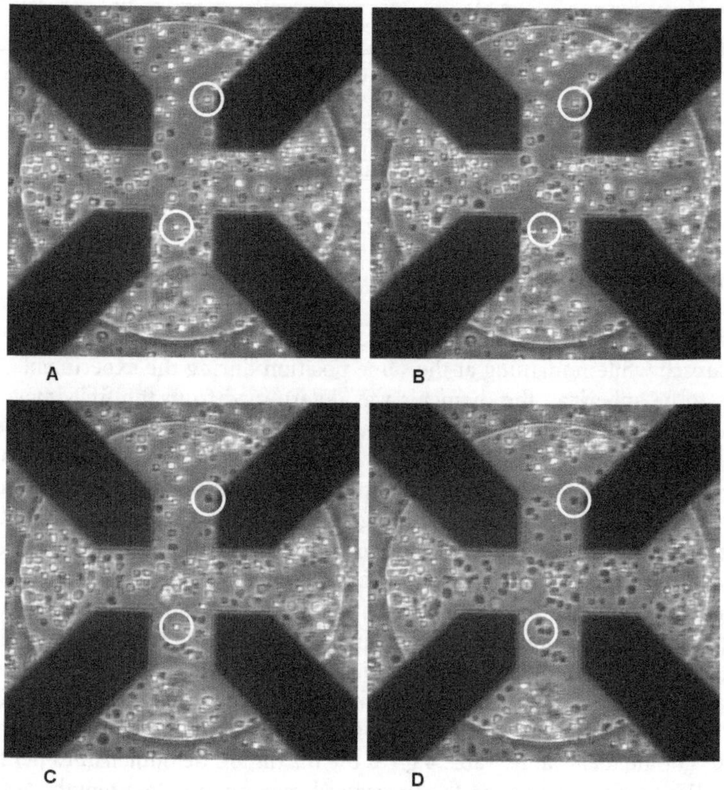

Fig. 24 Experimental results: situations at the start of the experiment [**A** (image0)], **B** 7 V_{pp}, **C** 12 V_{pp}, and **D** 17 V_{pp}. The frequency was set at 800 kHz. A *white circle* indicates two cells. At 12 V_{pp} the cell near the electrode (*upper circle*) has already collapsed; at 17 V_{pp} the other cell has also turned dark

Cells were positioned randomly inside the active region. However, this random positioning of the ten cells under study has to be comparable for the four experiments. Figure 26 therefore shows the average distance of the membrane sites of the cells nearest to an electrode edge and the electrode edge as a function of frequency. No significant differences were found, and thus the experiments were performed under comparable conditions.

From theory it may be expected that membrane breakdown leading to membrane rupture is dependent on a combination of the frequency of the field, temperature, the porated surface area of the membrane, and mechanical stress. Most of these parameters are dependent on the position of the cell inside a nonuniform electric field.

From the graph of Fig. 25 it is clear that a frequency dependency was present. This frequency dependency could be a direct result of the membrane area affected by electroporation. If a certain area needed to be electroporated before membrane rupture was induced, much higher field strengths were needed at higher frequencies (Fig. 23).

The results for 100 kHz show that membrane rupture occurred at a membrane potential of about 0.4 V for all cells. Therefore, at 100 kHz membrane breakdown was

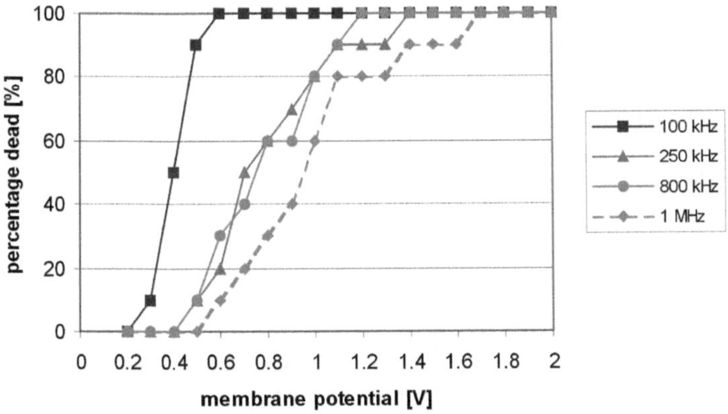

Fig. 25 Percentage of collapsed cells as a function of the induced membrane potential. Each graph consists of ten cells with varying diameter and located at different positions inside the field (Heida et al. 2002b)

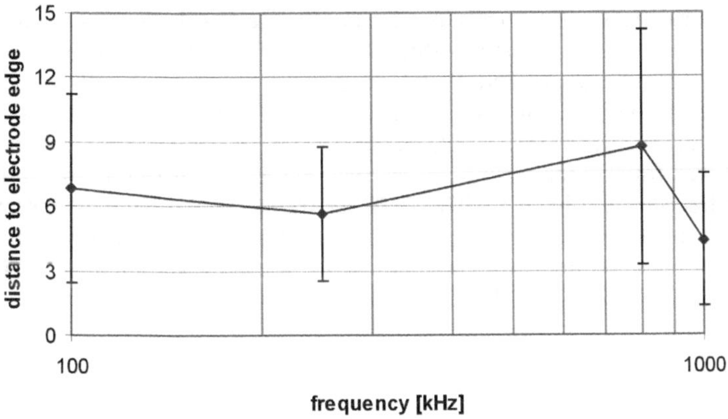

Fig. 26 Average distances between the membrane site of the cells nearest to an electrode edge and the electrode edge as a function of frequency

nearly independent of the position and radius of the cell and thus of the area involved in electroporation. Fig. 23A versus 23B confirms that the position of the cell does not affect the area of membrane breakdown at the lower frequencies (10 and 100 kHz) for a breakdown level up to about 0.5 V: the upper two curves (10 and 100 kHz) in both figures are almost identical.

Furthermore, when breakdown occurs at higher frequencies (>1 MHz) cell position starts to affect the membrane area involved in breakdown. This position-dependent breakdown level may be the cause of the decrease in the steepness of the slopes of Fig. 25 with increasing frequency. It has to be noted that the two situations shown in Fig. 21 do not cover the complete spectrum of possible positions. They represent "worst case" situations. It is, however, assumed that the relationship between electro-

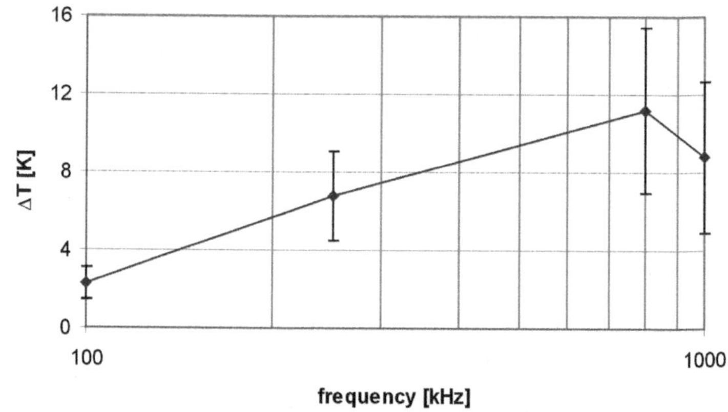

Fig. 27 Estimated average temperature rise for each frequency

porated membrane area and membrane potential will for most cells comply with the average of the curves of Fig. 25.

Mechanical stress on the cells arises from the electrical compressive forces on the membrane or from the pressure within ion channels to expand them, as described in Sect. 3.1.4 (Zimmermann and Neil 1996). These forces are hardly observable and were therefore omitted for discussion. Another mechanical force is created by the fluid flow arising from local Joule heating of the medium. From previous experiments it was found that a temperature rise of 1.5°C occurred at an input signal of 5 V_{pp} at frequencies above 1 MHz. At lower frequencies smaller temperature variations were measured which can be ascribed to the reduced field strength as a result of the electrode–medium interface impedance. A direct relation exists between temperature rise and input signal (Glasser et al. 1999):

$$\frac{\Delta T_1}{\Delta T_2} = \frac{V_1^2}{V_2^2} \tag{13}$$

For all four experiments an average input signal was determined at which breakdown occurred, incorporating the effect of the electrode–medium interface (Table 4), from which the average temperature rise was estimated. The result is shown in Fig. 27.

From Fig. 27 it may be expected that at the higher frequencies the mechanical force due the fluid flow exerted on a cell may lower its breakdown level; the larger the temperature rise the larger the mechanical stress exerted on a cell by the created fluid flow. In addition, with increasing temperature the membrane breakdown level generally decreases (Benz 1980a; Zimmermann and Neil 1996). This might explain the overlapping breakdown levels for 250 and 800 kHz in Fig. 25. With increasing frequency an increase in breakdown level is expected. However, due to the temperature rise the breakdown level decreases, which might be the case for the experiment performed at 800 kHz.

It may be concluded that the experimental results for nonuniform electric fields fit previous findings from literature for uniform fields (Benz 1980a, b, c; Zimmermann and Neil 1996; DeBruin 1999). In comparison to uniform fields, field frequency also greatly determines the breakdown level of the cell membranes in nonuniform fields. In addition, with increasing temperature the breakdown level was found to decrease.

4 Investigating Viability of Dielectrophoretically Trapped Neuronal Cells

Trapping neurons by negative dielectrophoretic forces means that they are repelled from the regions of high electric field strength. It may therefore be expected that, even though membrane breakdown may occur at these high field strengths, cells do not necessarily have to be damaged irreversibly while being trapped.

The investigation of the viability of trapped cells was divided into two experiments depending on the frequencies used. First, a high frequency was used (14 MHz) at which the induced membrane potential is low even at a high field strength. The morphology of the neuronal cells was analyzed. Secondly, viability was investigated for the more critical frequency range from 10 kHz to 1 MHz. The trapped cells in this case were analyzed using a staining method.

4.1
Viability of Neuronal Cells Trapped at a High Frequency

4.1.1
Experimental Procedure

4.1.1.1
Experimental Setup

The surface of the electrode plate inside the culture chamber was coated with PolyEthylenImine (PEI, 30 ng/ml; Sigma-Aldrich) before the experiment was started. PEI has been shown to support attachment and growth of neurons comparable to polyornithine and polylysine (Rüegg and Hefti 1984). Due to the strong interaction between cells and the PEI surface coating, cell clustering can be prevented and monolayer cultures of neurons created. Moreover, cell adhesion mediates morphological changes, growth, and maturation (Lelong et al. 1992).

Directly after dissociation of the cortical cells the experiments were started. A drop of cell suspension (20 μl, 0.7×10^6 viable cells/ml) was pipetted on top of each of the four electrode structures (only electrode structures with triangular tips were used) as well as a reference region on the electrode plate. The cells that were not subjected to the electric field (the control cells) were thus cultured on the same electrode plate as the trapped cells, and were therefore exposed to the same culturing conditions.

The culture chamber was sealed using a Petri dish that enclosed the glass ring neatly. The electrode plate was clamped into a small box in which small pins provide the connection from the gold electrodes to the coaxial cable leading to the signal generator. This box was directly placed on a Nikon Inverted microscope (Diaphot-TMD) where the field was applied for 1 h (3 V_{pp}, 14 MHz, room temperature, minimal light exposure). In total five electrode plates were used subsequently during an experimental session (5 h total). During this period the remaining dissociated cortical cells were kept in a plastic tube filled with R12 medium at room temperature to slow down physiological processes. The tube was regularly shaken to prevent the cells from adhering to the inside of the tube. A total of six experiments were used for the analysis.

Images were taken from the four electrode structures directly after the field was switched off by a digital video camera (Sony CCD Camera, Model DXC-151P) to determine if the cells were really dielectrophoretically trapped before continuing the analysis. These images, however, were not ideal for accurate investigation due to poor phase contrast as a result of the curving surface of the drop of cell suspension.

In about 7 min all cells were precipitated onto the PEI-coated substrate after the application of the drops of cell suspension. In 1 h the cells adhered to the PEI substrate and the electrode plate could be removed from the box without repositioning the cells. The Petri dish sealing the culture chamber was replaced by the lid of a Petri dish so that it was not completely sealed off anymore. The electrode plate was stored in the incubator (37°C, 5% CO_2, 100% humidity) positioned a few meters away from the microscope. After 3 h, 2 ml of fresh R12 medium was applied to the culture chamber and again images were taken using the microscope and camera as mentioned above. On the first day in vitro (DIV) three photo sessions with intervals of 4 h were carried out. Thereafter images were taken once a day. Medium was refreshed after 3 days.

Postnatal cortical cells are generally more vulnerable to culturing conditions than fetal cells, and differences can be detected in dendritic outgrowth and the rate of axonal elongation (Banker and Goslim 1998). Furthermore, during the first hour of the experiment and the time periods in which images were taken (about 30 min for each photo session) the cultures were not kept under optimal conditions, which would be room temperature (21°C) in the absence of 5% CO_2, and exposure to light (Banker and Goslim 1998). Thus, the experimental conditions lead to a loss of viability extraneous to field exposure.

4.1.1.2
Data Analysis

Cell dimensions and the length of the processes were determined using the software package VIDAS (Carl Zeiss B.V., Weesp, The Netherlands). By drawing the contours around the soma of the cells the areas were automatically calculated. The number of processes originating from the neuron as well as their branches were counted manually.

The area within which the images were analyzed was restricted to the square area of 5,000 μm^2 enclosed by the four electrode tips, termed the center. For the control experiments a randomly chosen square area of the same size was used. Cells were allowed to partially cross two of the sides of the square area. The processes were allowed to cross all sides of the area when originating from cells that satisfied the previous constraint. In case of branching processes, the length of the longest branch was taken to be that of the "original" process (no area restriction applied). The results from six experiments were averaged at eight time points for trapped and reference cortical cells up to 5 days.

4.1.2
Experimental Results

4.1.2.1
Number of Outgrowing and Nonoutgrowing Cortical Cells

Image analysis showed that cells that initially occupied the defined area kept to their initial position quite closely for some days. The images of Fig. 28 show the neural

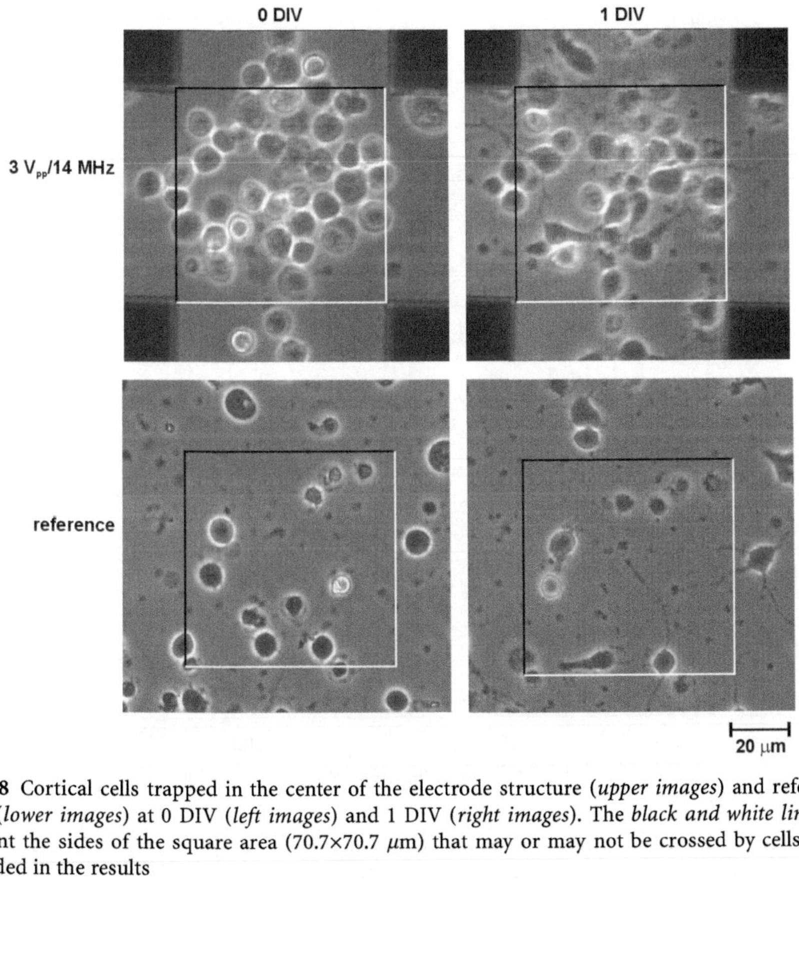

Fig. 28 Cortical cells trapped in the center of the electrode structure (*upper images*) and reference cells (*lower images*) at 0 DIV (*left images*) and 1 DIV (*right images*). The *black and white lines* represent the sides of the square area (70.7×70.7 μm) that may or may not be crossed by cells to be included in the results

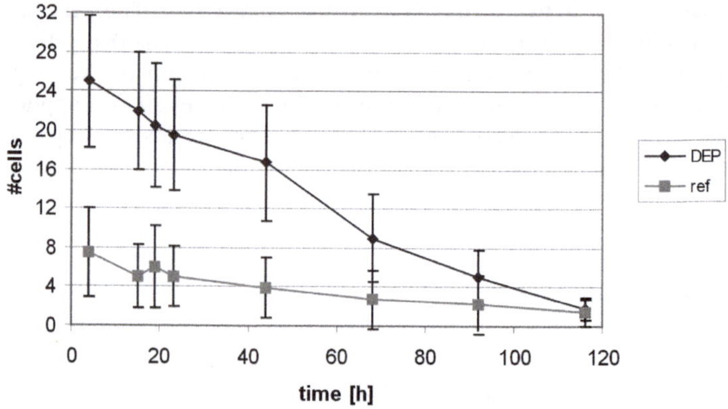

Fig. 29 Total number of cortical cells within a defined area for trapped (*DEP*) as well as reference (*ref*) cortical cells

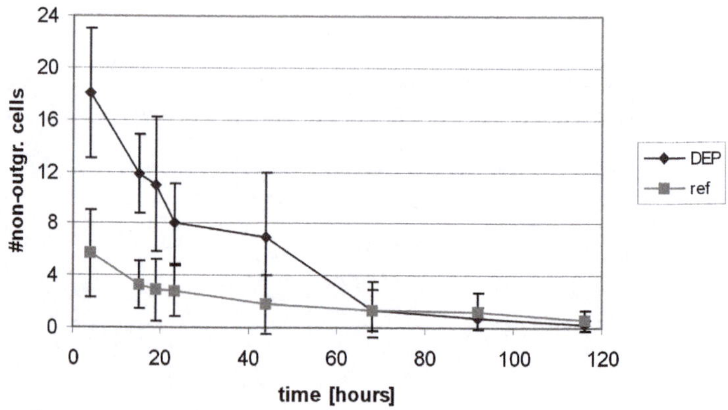

A

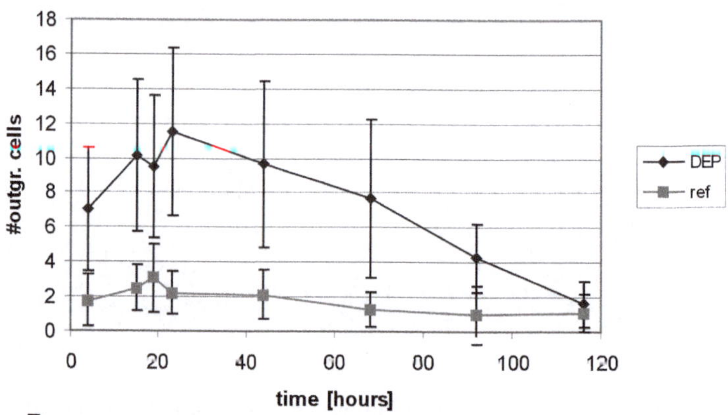

B

Fig. 30 A Number of nonoutgrowing trapped (*DEP*) and reference (*ref*) cortical cells. **B** Number of outgrowing trapped (*DEP*) and reference (*ref*) cortical cells cultures in the center of the electrode structure and a reference area after 0 and 1 DIV (Heida et al. 2001b). Inherent in the process of trapping, the number of DEP-trapped cortical cells was always larger than the number of reference cortical cells within an area of the same size. Fig. 29 shows the average number of trapped and reference cortical cells during the experiments. Both curves show a decreasing trend with almost no cortical cells left after 5 DIV.

The total number of cells is the sum of nonoutgrowing and outgrowing cells. Figure 30 shows the number of nonoutgrowing and outgrowing cells over time. An average of 25 cells were dielectrophoretically trapped and about half of them grew out. In the reference situation an average of 7 cells were initially located within an area of same size. Of these cells, an average of 3 grew out. Both curves of Fig. 29 end at an average of 3 cells at 5 DIV of which 2 cells still showed processes.

Processes started to grow out during the first 24 h. It appears that cells that were seen as nonoutgrowing cells at 3 h, grew out afterwards. Thus, the decrease in the number of nonoutgrowing cells was accompanied by an increase in the number of outgrowing cells during the initial stage of the experiment (Fig. 30). Both curves show a decrease after 24 h as a result of the natural degeneration process.

4.1.2.2
Area of the Cortical Cells

The average area taken up by the soma of outgrowing cells increased from 70 to 95 μm^2 for trapped as well as reference cells. The area of nonoutgrowing cells varied around 50 μm^2. Figure 31 shows the results as a function of time. There is no significant difference for the average area between reference and trapped cortical cells both for outgrowing and nonoutgrowing cells.

4.1.2.3
Number of Processes

Analogous to the number of outgrowing cells, the average number of processes within the DEP area was larger compared to that within the same-sized reference area up to 5 DIV as shown in Fig. 32. The number of processes after 5 DIV was comparable for both situations. Per outgrowing neuron the average number of processes varied between 1 and 2 during the experiment for both situations.

4.1.2.4
Process Length

The cells started to grow out after several hours of culturing. The length of the processes was rather variable as shown by the large standard deviations in Fig. 33. From this graph no significant difference in the length of the processes over time between the two types of cells, trapped and reference, could be detected. The maximum length measured was around 30 μm for both types after 5 days in vitro.

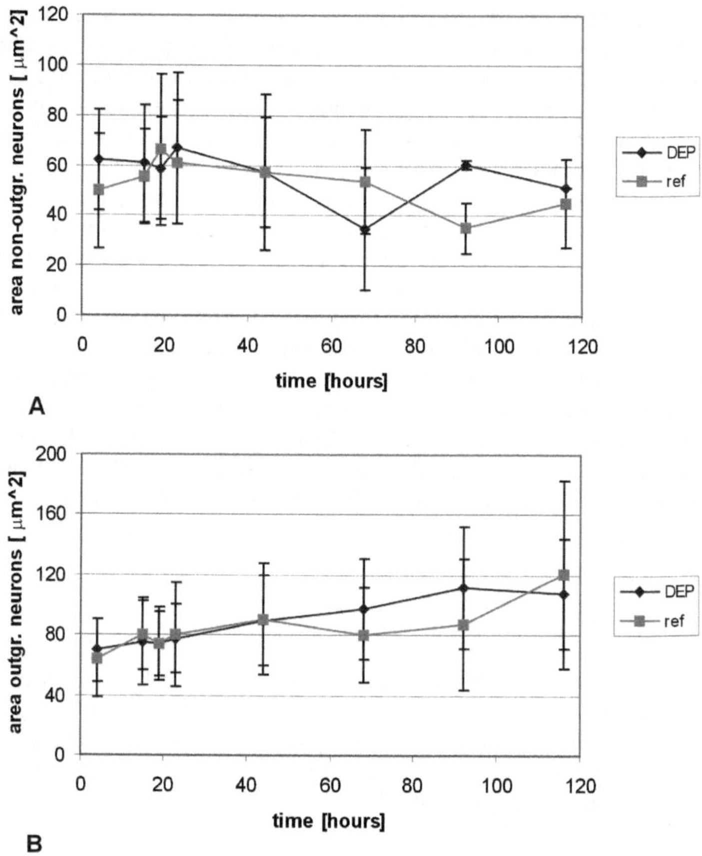

Fig. 31 **A** Average area taken up by the soma of nonoutgrowing cortical cells during the experiment for both situations (*DEP* and *ref*). **B** Average area taken up by the soma of outgrowing cortical cells during the experiment for both situations (*DEP* and *ref*)

4.1.2.5
Data Comparison

It was expected that if the DEP procedure would influence the viability and adhesiveness of the cortical cells, this effect would have been seen during the initial stage of the experiments, i.e., directly after electric field application.

The size of the cells were comparable for both situations, meaning that no (temporary) membrane breakdown resulting in osmotic events or changes in membrane properties had occurred. In addition, no differences in the length or number of outgrowing processes per cell were detected. These parameters were equal for the trapped and reference situation in spite of the different number of cells located within the defined area at the onset of the experiment.

In order to be able to compare the curves showing the number of cells (Figs. 29 and 30) for the trapped and reference situation, the initial condition should be the

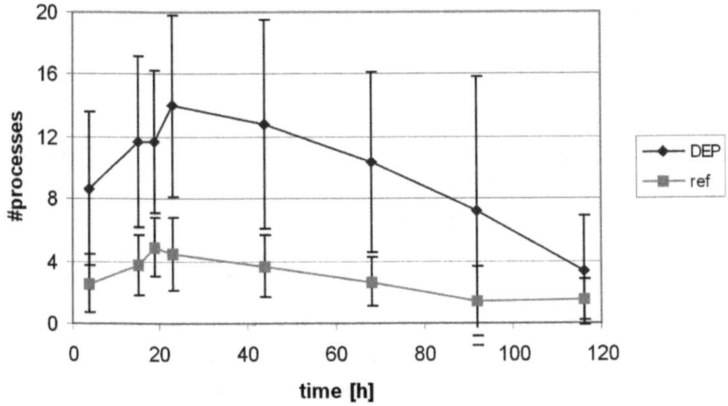

Fig. 32 Number of processes as a function of time for trapped (*DEP*) and reference (*ref*) cortical cells

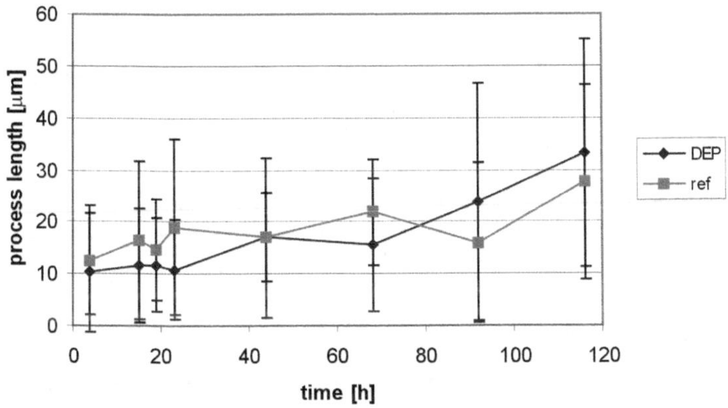

Fig. 33 Length of the processes as a function of time for trapped (*DEP*) and reference (*ref*) cortical cells

same. This can be accomplished by normalization of the data, i.e., dividing all data points by the value of the first data point. Fig. 34A shows the result for the total number of cells. No significant differences between the data points of the two curves are shown here, and thus the differences in Fig. 29 are eliminated. The same was true for outgrowing and nonoutgrowing cortical cells (Fig. 30). For comparison of the number of processes (Fig. 32) the same method can be applied. The result is shown in Fig. 34B. Again no significant differences between the two curves are present. The decrease in the total number of cells within the defined area can therefore be ascribed to the experimental conditions in both cases. Thus, the viability of the trapped cortical cells was not influenced by the electric field (3 V_{pp}/14 MHz). This does, however, not automatically mean that the field had no influence on cell functioning. This was investigated in more detail and is described in Sect. 4.3.

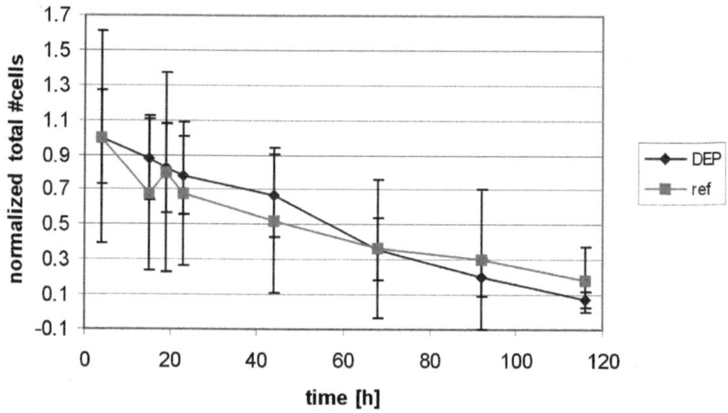

A

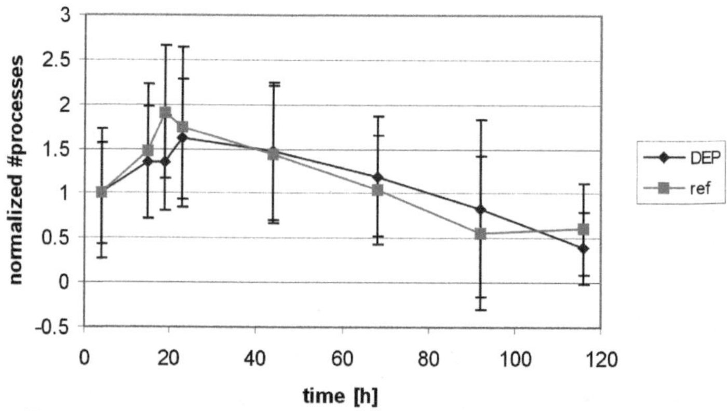

B

Fig. 34 **A** Normalized total number of cortical cells for the trapped (*DEP*) and reference (*ref*) situation. **B** Normalized number of processes for both situations

4.2
Viability of Neuronal Cells Trapped at Low Frequencies

4.2.1
Theoretical Estimation of the Maximum Membrane Potential

The electric field was computed in Sect. 2 for both quadrupole electrode structures. The maximum field strengths were 108.65 and 181.09 kV/m for the triangular electrode tips for 3 and 5 V_{pp}, respectively; for the circular electrode tips the maximum field strengths were 121.7 and 202.93 kV/m, respectively. In practice, these field strengths could not be realized due to less than ideal material properties.

Table 5 lists the estimation of the reduction in the electric field strength together with the accompanying maximum induced membrane potentials $V_{m,max}$ as calculated according to Eq. 10, for input signals of 3 and 5 V_{pp}. At the lower frequencies the estimated maximum induced membrane potential is smaller as compared to that at

Table 5 Percentage of the field strength in relation to the theoretical field strength[a]

Frequency [kHz]	$\frac{E_{estimated}}{E_{theoretical}} * 100\%$	$V_{m,max}$ [V] (3 V_{PP}) tri	cir	$V_{m,max}$ [V] (5 V_{PP}) tri	cir
10	44%	0.36	0.40	0.60	0.67
100	44%	0.36	0.40	0.60	0.67
1,000	88%	0.65	0.73	1.08	1.21
14,000	90%	0.11	0.12	0.19	0.21

[a] Cell radius r=5 μm, membrane capacitance C_{mem}=9 mF/m^2, cytoplasmic resistivity ρ_{int}=1.33 Ωm, and medium resistivity ρ_{med}=0.625 Ωm

1 MHz, since the electric field strength is reduced due to the electrode–medium interface impedance. At 14 MHz the capacitive component of the membrane acts as a "short cut" and thus induced membrane potentials are expected to be small.

4.2.2 Experimental Procedure

4.2.2.1 Experimental Setup

The surface of the electrode plate inside the culture chamber was coated with PolyEthylenImine (PEI; Sigma-Aldrich), 30 ng/ml, before the experiment was started. Electrode structures with both types of tips (semi-circular and triangular) were used.

Immediately after dissociation of the cortical cells the experiments were started. A drop of cell suspension (20 μl, 1.5×10^6 viable cells/ml) was pipetted on top of each of the four electrode structures as well as on a reference region on the electrode plate.

The culture chamber was sealed off and the electrode plate was then clamped into the small box in which tiny pins provide the connection to the signal generator. This box was directly placed on the microscope (Diaphot-TMD) where the field was switched on (t=0). It was applied for 30 min (room temperature, minimal light exposure). Sinusoidal input signals of 3 and 5 V_{PP} at 10, 100 kHz, and 1 MHz were used. Extra experiments were performed at 3 V_{PP}/14 MHz.

In about 15 min all cells were precipitated and positioned (by dielectrophoretic forces) onto the PEI-coated substrate. After 30 min the field was switched off and images were taken from the eight electrode structures by a digital video camera (Sony Model DXC-151P). The electrode plate was removed from the box and was stored in the incubator (37°C, 5% CO_2, 100% humidity). After 2 h and 30 min, 2 ml of fresh R12 medium was applied to the culture chamber and again images were taken.

On the first, second, or fifth DIV cell death or viability was investigated using a staining procedure. Acridine Orange and Propidium Iodide staining were used for staining the cells [10 ml PI (40 μg/ml) + 14 phosphate buffered saline (PBS) + 24 μl AO (5 μg/ml)].

Acridine Orange is a weak base that enters the cells to function as an inclusion dye and cause viable cells to fluoresce green. Propidium Iodide, an analog of ethidium bromide, is as an exclusion dye that cannot penetrate living cells but is able to enter dead or dying cells, bind to nucleic acids, and cause them to fluoresce bright red (Banks 1988).

Digital images were taken using fluorescence as well as phase contrast microscopy. After staining, the cultures were terminated. In total six experiments were used for the analysis of each of the three situations (1, 2, and 5 DIV).

4.2.2.2 Data Analysis

Image analysis was performed to determine the following parameters:

- The total area covered with cells at 30 min
- The area taken up by individual cells at 3 h (after applying fresh medium)
- The area of green and red stained cells at 1, 2, and 5 DIV

4.2.2.3 Contour Detection for Area Determination

The total area covered with cells in the center of the electrode structure at 30 min was determined by the contours of cells and/or groups of cells. The area taken up by individual cells at 3 h was determined by the same method. A Matlab script (Matlab 5.3, The MathWorks, Inc., Natick, MA, USA) was developed which started with the detection of the center area between the electrode tips. It was assumed that the rotation of the electrode structure was minimal, and a square area 71×71 μm indicated the center area within which the analysis was performed. For cells crossing the boundaries of this area, only those parts located inside were included in the total cell-covered area. An example is given in Fig. 35 where the square center area is indicated as well as the contour of two cells.

Except for the detection of the electrode tips, the analysis for the reference situation was similar to that described above.

4.2.2.4 Detection of Red and Green Stained Areas

From the phase-contrast images the position of the center area between the electrodes was determined. After enhancing the contrast of the fluorescence image, the total area of red and green within this center area was determined using the same coordinates for the center. Contrast enhancement was used to make the division between the green and red areas more distinct. Parts of processes that were vaguely stained were removed and did not affect the results. The images were RGB bitmaps, which means that the image was an array of (m×nx3), i.e., an array of (mxn) pixels for each color (red, green, and blue). Overlapping areas of green and red appeared as yellow. These areas were categorized as being red and were therefore subtracted from the green area. From the number of red and green pixels the red and green areas were calculated.

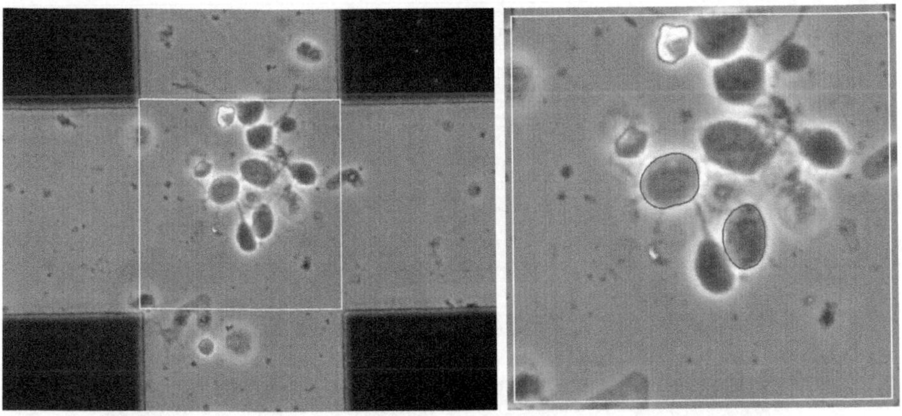

Fig. 35 **A** Determination of the center area (the small rotation of the electrode structure was neglected). The image shows the situation after 3 h (3 V_{pp}/14 MHz). **B** Cell areas were calculated to be 73.7×10^{-12} m^2 for the cell on the *right* and 86.4×10^{-12} m^2 for the cell on the *left*

The same procedure was used for the reference situation, where a same-sized square area was taken within which the green and red stained areas were determined.

4.2.3 Experimental Results

4.2.3.1 Total Area Covered with Cells

Due to differences in electric field and frequency-dependent dielectrophoretic forces (Heida et al. 2001a), the number of cells initially trapped was different for each field setting. The total area covered with cells after 30 min of field application varied between 10^{-9} m^2 at 3 V_{pp}/10 kHz to about 5×10^{-9} m^2 at 5 V_{pp}/10 kHz. Figure 36 shows the area in question after 30 min of field application for four different situations.

After 3 h fresh medium was applied to the cultures. Due to the mechanical force on the cells while new medium was applied, cells that had not adhered (well enough) to the substrate were washed away. Figure 37 shows the area covered with cells after 3 h as a percentage of the initial situation (at 30 min). This graph was determined by averaging data from different experiments (n=6) that were performed with identical field settings. It shows that, especially at 5 V_{pp}/10 kHz, many cells did not adhere before medium was applied and were thus washed away (at 3 h). Only in the reference situation and at 3 V_{pp}/10 kHz a maximum of about 80% of the initial area remained covered, while at 5 V_{pp}/10 kHz the smallest percentage of remaining cells was found. At 100 kHz and 1 MHz almost no significant difference was detected between the two input values. For the 5 V_{pp} situation a slight increase in percentage could be seen with increasing frequency from 10 kHz up to 1 MHz.

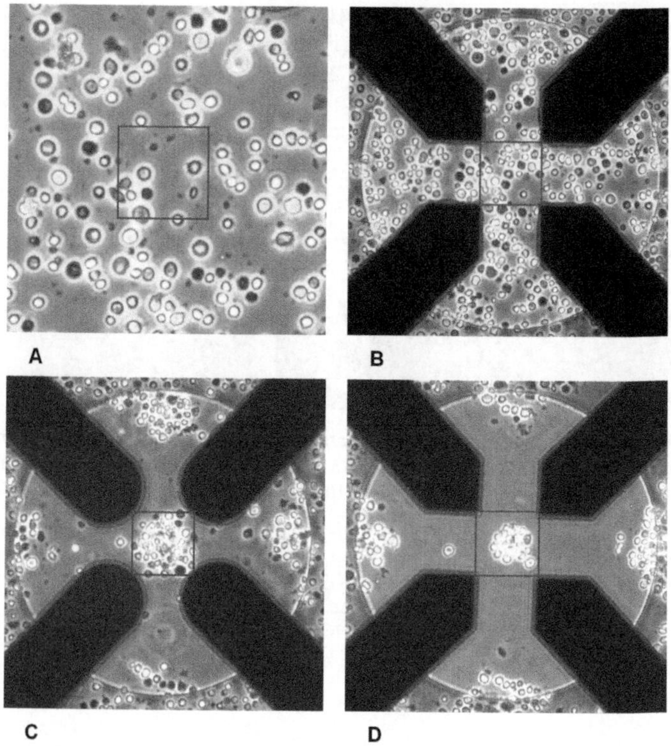

Fig. 36 Coverage of the center with cells after 30 min of field application for **A** reference situation, **B** 3 V_{pp}/10 kHz, **C** 3 V_{pp}/1 MHz, **D** 5 V_{pp}/1 MHz

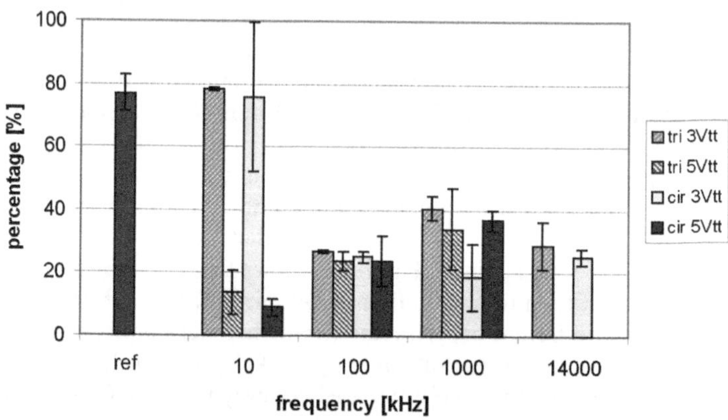

Fig. 37 Total area covered with cells at t=3 h as a percentage of the area covered after 30 min of field application. The *leftmost bar* represents the reference situation

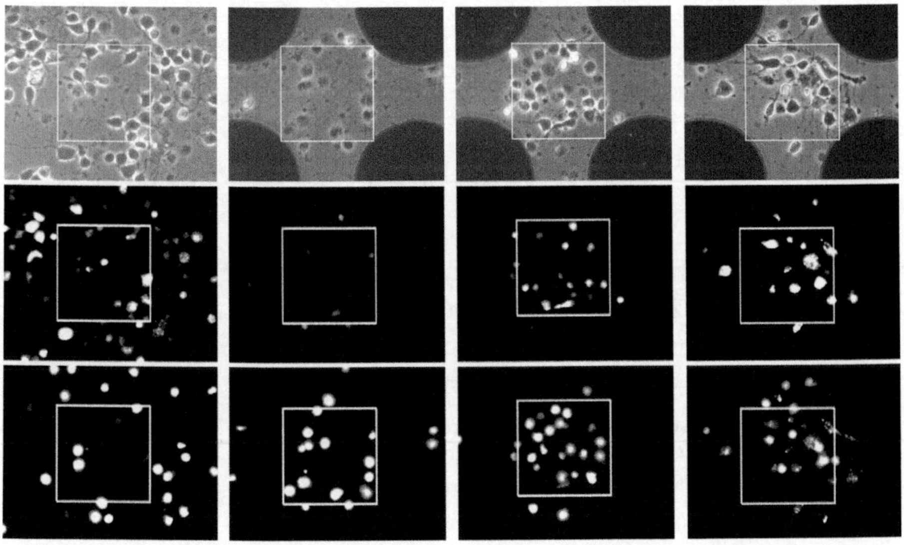

Fig. 38 Cell cultures after 1 DIV for **A** the reference situation, **B** 3 V_{pp}/10 kHz, **C** 3 V_{pp}/100 kHz, and **D** 3 V_{pp}/1 MHz. The *upper images* are phase contrast images; the *middle level* shows the green channel of the images that were taken using fluorescence microscopy; the *lower level* shows the red channel of these fluorescence images. (PI stains dead cells red, and viable cells are stained green by AO)

From the images at 3 h, the area of the individual cells was found to be around 70×10^{-12} m^2. No significant differences were found for the different situations, indicating that no cell swelling due to field application could be detected after 3 h of culturing. This result is in accordance with the results found previously for the reference and 3 V_{pp}/14 MHz situation (Heida et al. 2001b). It confirms that the osmolarity of the R12 medium resembles that of the interior of the neuronal cells (Romijn et al. 1984).

4.2.3.2
Staining of DEP-Trapped Cells

Figure 38 shows the results after 1 DIV for the reference situation and the circular geometry for 3 V_{pp} at 10 kHz, 100 kHz, and 1 MHz. The graphs in Figs. 39, 40, and 41 show the average of the total areas covered with cells that were colored green and red inside the center. Two bars represent one situation of which the left bar indicates the green area and the right bar the red area. Only the graphs of the circular electrode structure are shown. The graphs of the triangular geometry differed somewhat in absolute values, however, the trends were similar to those of the circular geometry.

The cell suspensions used for the staining experiments at different DIVs did not come from the same batch. Therefore, differences in the number of viable cells at the start of the experiment in combination with small variations in fluorescence intensities resulted in the large standard deviations. This makes it possible that at 5 DIV the

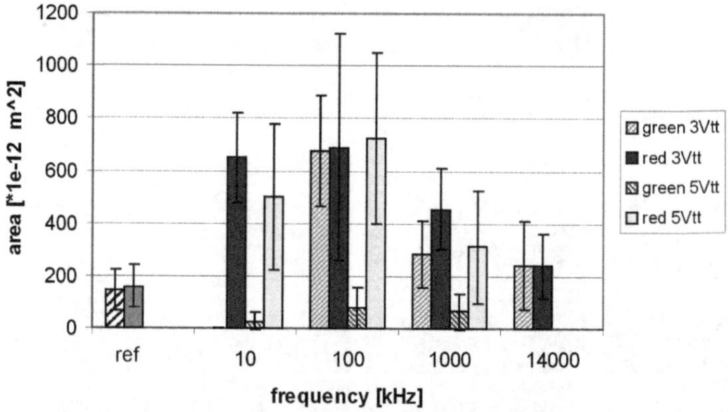

Fig. 39 Green and red stained areas for the different situations after 1 DIV (semicircular geometry). The *two bars* on the *left* indicate the reference situation: the *left* and *right bars* represents the green and red stained area, respectively

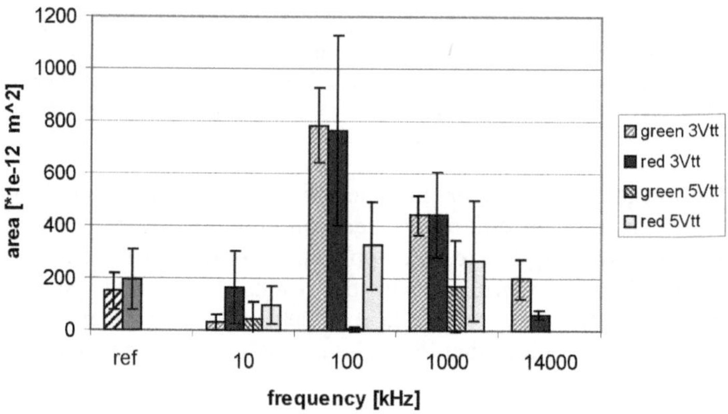

Fig. 40 Green and red stained areas for the different situations after 2 DIV (semicircular geometry). The *two bars* on the *left* indicate the reference situation: the *left* and *right bars* represents the green and red stained area, respectively

total area was somewhat larger than at 2 DIV or even 1 DIV (e.g., compare the situation at 3 V_{pp}/1 MHz for all DIVs).

According to Figs. 38–41, the reference situation and experiments performed at 1 MHz and 14 MHz show an almost constant total area up to 5 DIV. At 3 V_{pp}/10 kHz many cells maintained adherence inside the center after the application of fresh medium. However, most cells appeared to be dead after 1 DIV and left the area after 2 DIV. At 5 V_{pp}/10 kHz the survival chance of dielectrophoretically trapped cells was minimal, which was already noticeable after 1 DIV. At 100 kHz survival chances were higher at 3 V_{pp} than at 5 V_{pp}.

When assuming a frequency-independent level of the membrane potential above which electroporation occurs, the experimental results contradict the estimated max-

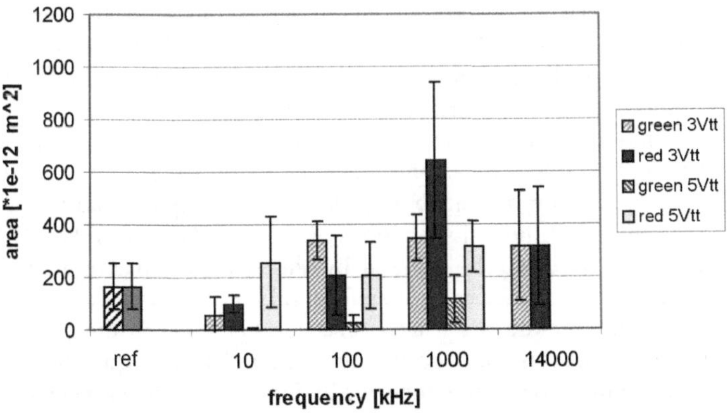

Fig. 41 Green and red stained areas for the different situations after 5 DIV (circular geometry). The *two bars* on the *left* indicate the reference situation: the *left and right bars* represents the green and red stained area, respectively

imum induced membrane potentials (Table 5). Fewer viable cells would have been expected at 1 MHz in comparison to the other frequencies, but Figs. 39–41 clearly show that survival chances were smallest at 10 and 100 kHz, especially at 5 V_{pp}. In addition, the results for the semicircular and triangular electrode structures showed the same trend, while the absolute values of electric field strength and maximum induced membrane potential differed slightly (Table 5). Membrane breakdown leading to cell death is thus mainly determined by the frequency of the field. At 10 kHz, however, Fig. 37 shows a strong amplitude-dependent behavior, not seen for the other frequencies. This graph clearly shows that the field had affected adhesive properties of cells. Except at 10 kHz this process was found to be slightly dependent on frequency, but almost independent of amplitude. In explaining the results, two effects need to be focused on: (1) a frequency-dependent survival chance, and (2) frequency- and amplitude-dependent changes in adhesive properties.

4.2.3.3
Adhesion in Relation to Field Strength and Frequency

Changes in membrane and/or cell properties due to reversible or irreversible membrane breakdown may lead to changes in adhesion. The curve giving the area covered with cells after 3 h of incubation as a percentage of the area taken up by the dielectrophoretically trapped cells indicates that for all field settings adhesive properties may have been changed or lost during field exposure, except at 3 V_{pp}/10 kHz.

Use of a single-shell model implies an assumption that the dielectric properties of the cell membrane are constant. Since the induced membrane potential is radius-dependent according to Eq. 1, the cell organelles will in general not directly be affected by the electroporation of the outer membrane. This permits the use of a single-shell model with the inherent assumption that viability is directly related to the breakdown of the cell membrane.

However, many relaxation processes were found to occur in the membrane in the frequency range up to 100 kHz, referred to as α- and β-dispersions (Schwan 1957; Pethig 1990). In the β-dispersion range (1 kHz–100 MHz) the main components are Maxwell–Wagner interfacial polarization mechanisms caused at the interfaces of two media with different dielectric constants and conductivities (Schwan 1957; Pohl 1978; Chiabrera et al. 1985; Gimsa and Wachner 1998). Other processes such as surface conductance, membrane transport processes, and dipole relaxations of macromolecules may influence these dispersions (Gimsa and Wachner 1998). Nevertheless, these dispersions may easily be masked by the stronger structural β-dispersions. Still, the complex capacitance may therefore vary not only because of changes in membrane area and thickness, but also because of changes in dielectric increments due to relaxation processes, each occurring at a particular frequency (Awayda et al. 1999). The membrane capacitance can thus be modeled as the parallel sum of capacitive increments, each with its own time constant. Because the cell membrane contains a large variety of channels, transporters and other proteins, it seems likely that the low-frequency dispersions may arise from dipoles associated with integral membrane-spanning proteins. In addition, charge movements within the membrane associated with the gating mechanism of excitable channels in nerve membranes may be responsible for dielectric dispersions (Awayda et al. 1999).

Studies can be found in literature on the reorganization of the microfilament structure induced by alternating electric fields (Cho et al. 1996). Microfilaments are important for cell proliferation, differentiation, activation, locomotion, phagocytosis, and attachment. It was postulated that the reorganization of cell-surface receptors was caused by reorganization of cytoskeletal structures in response to the field, since transmembrane proteins interact with cytoskeletal elements (Cho et al. 1996). The effect was critically dependent on the frequency of the field; microfilaments were unable to reorganize in response to more rapidly oscillating electric fields. However, microfilament reorganization was also found to be dependent on the strength of the electric field; it increased nonlinearly with the field strength. Frequencies up to 60 Hz and field strengths up to about 2 kV/m, which did not result in membrane breakdown, were used. These parameters are far out of our range of field parameters, but the existence of a field-strength-dependent process that might affect cell adhesion has indeed been proven. The exact physical processes occurring in the frequency range around 10 kHz are yet unknown.

Even though at 3 V_{pp}/10 kHz the adhesive properties of cells apparently were still intact after 3 h (Fig. 37), membrane breakdown had probably occurred. The cells were not able to recover because they were dead after 1 day in vitro (Fig. 39).

4.2.3.4
Viability in Relation to Frequency

The pulse-length dependence, as described in Sect. 3.1.2, indicates that the breakdown potential increases with decreasing pulse length. In the experimental situation described here this effect may be translated into a frequency dependency, i.e., the membrane is able to withstand larger field strengths at higher frequencies as compared to lower frequencies (see Figs. 39–41). Possibly the membrane is able to recover more rapidly from a briefly applied electric field pulse (i.e., high frequency) than from a long pulse. In literature it was already noted that repeated pulsing of the membrane with pulses of long duration resulted in an increase in permeability with each subsequent pulse, due to which the membrane finally was seen to lose its ability to reseal (Ryttsén 2000; Zimmermann and Neil 1996). This effect can be recognized when comparing the results of the 10 kHz experiments with the 100 kHz results, both at 3 V_{pp}: at 10 kHz cells were dead at 1 DIV, while at 100 kHz they were still alive (Fig. 39). At between 2 and 5 DIV, however, a large decrease in the total area covered was seen at 100 kHz. At 5 V_{pp} this decrease has already started after 1 DIV, indicating that an increase in amplitude is comparable to a decrease in frequency. Similar effects were detected by (Ryttsén 2000): for higher potentials the high conductance state lasted longer than for near-threshold potentials. Furthermore, the corresponding membrane ion permeability was higher. These two effects might induce earlier cell death as compared to lower potentials and/or higher frequencies.

In literature a theory was presented that explains the pulse length dependence, which may also be applicable to the presented situation. Assuming that the membrane can be regarded as a capacitor filled with a homogeneous elastic material, it was postulated that the thickness of the membrane capacitor depends on the electrical compressive and mechanical forces (Sukhorukov et al. 1998; Zimmermann and Neil 1996). If the membrane material is not perfectly elastic and shows inertia, the elastic compressive modulus perpendicular to the membrane plane becomes a function of compression time. In addition, it increases with increasing rate of compression, comparable to visco-elastic materials.

4.2.3.5
Cell Death

The region of maximum field strength is small in comparison to the cell dimensions. Field gradients are largest in the vicinity of the high-field-strength regions. The negative dielectrophoretic forces created on the cells in the regions of high field gradient direct the cells toward regions of minimum field strength. Due to the field nonuniformity and the movement of the cells, it becomes difficult to predict the fractional area affected by electroporation.

It may be expected that only a small part of the membrane on one side of the cell was electroporated. In addition, probably only a relatively small fraction of the total number of cells that were trapped in the center were exposed to the maximum field intensities. This leads to the conclusion that even a small fractional area of the membrane being electroporated was enough to cause cell death.

Breakdown levels could be estimated to be about 0.3 V at 10 kHz. At higher frequencies this level is expected to be about 0.6 V, according to Table 5.

Most probably, breakdown was irreversible at 10 kHz while at 100 kHz it may have been reversible. At 100 kHz 50% of the cells were still viable after 1 DIV, comparable to the reference and the 3 V_{pp}/14 MHz situation. After 1 DIV the total area covered with cells decreased significantly. A possible explanation for cell death in this case can be found in the constituents of the medium. In a recent publication it was declared that large concentrations of Na^+, a pH indicator (e.g., phenol red), and the use of organic buffer substances should be omitted because their (continuous) uptake by the cell may lead to adverse effects (Zimmermann and Neil 1996). The medium used in the experiments described in the present review is a chemically defined culture medium (Romijn et al. 1984), which contains a relatively large concentration of Na^+, the buffer HEPES, and phenol red. Cells could have died due to the uptake of medium constituents.

In conclusion it may be stated that higher survival rates may be expected when using medium consisting of ultrapure substances dissolved in double distilled water (Zimmermann and Neil 1996). However, avoiding membrane breakdown may be the best solution. Therefore, a frequency of 100 kHz is the lower limit at low amplitudes (3 V_{pp} in this case) and 1 MHz at higher amplitudes.

4.3
Recording Neuronal Activity

Cell viability was preserved when using frequencies above 100 kHz and electric fields of moderate field strength (<80 kV/m), as was shown above. At frequencies below 1 MHz membrane breakdown occurred at high field strengths and it was found that cells were generally not able to recover completely after the field was switched off. Although many cells remained viable during the first two days in vitro more than 50% died thereafter. Viability, however, as investigated by previously described methods does not imply that neuronal activity has not been altered by the field. In a pilot study a start has been made with the investigation of the spontaneous activity of DEP-trapped cortical rat neurons.

4.3.1
Extracellular Recording

The conventional method of observing the electrical signaling of active nerve cells uses microelectrodes in glass pipettes. This method is not ideal for neural networks for two reasons: the cells are traumatized by the insertion of the microelectrodes (which shortens their life span), and it is physically very difficult to insert many electrodes into the same preparation (Wilkinson 1995). So the idea of using structures that have incorporated into them extracellular, and so essentially passive, electrodes to monitor the nervous signaling, is very attractive. The MEA as presented in Sect. 1 can be used for the recording of neuronal activity.

Extracellular electrodes positioned close to firing neurons pick up the external ionic flux, the action potential, that flows out of the cell through the ion gates in the membrane of the cell. This ionic current is converted into a current carried by electrons in the electrode, which can be measured. However, when a metal is brought in contact with an electrolyte, an interface impedance is created as described in Sect. 2.

Recording, as well as stimulation, requires a tight contact between the neuron and the electrode (Bove et al. 1995; Buitenweg et al. 1998; Regehr et al. 1989). A schematic drawing of a neuron sealing an electrode is given in Fig. 42. Biological cells are very poor conductors at low frequencies (<10 kHz) and therefore they force electrical currents to bypass them. The resistance of the sealing gap determines these leakage currents. When the neuron covers the electrode completely this resistance is very high,

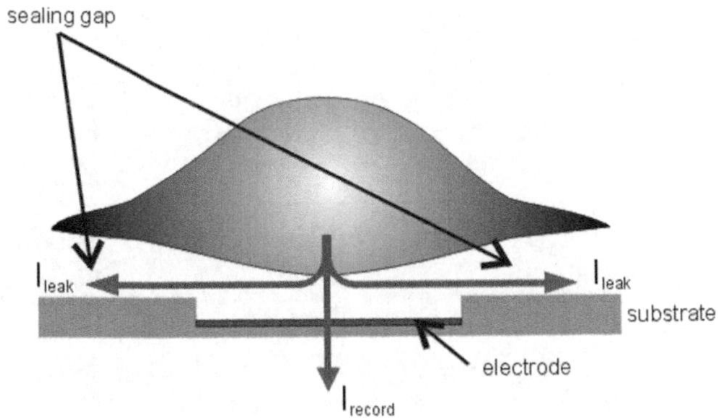

Fig. 42 Schematic drawing of a neuron positioned on top of an electrode. Without complete coverage of the electrode, currents may leak away via the sealing gap

with the result that the leakage current is very low. It was also found that the signal-to-noise ratio of the recordings increases with increasing sealing resistance (Buitenweg et al. 1998). The sealing resistance turned out to be a major factor in ensuring the neuron–electrode contact.

Important properties of extracellularly recorded spontaneous activity are the amplitude of the spikes, the signal-to-noise ratio, the regularity with which the spikes occur, and the shape of the signals. These parameters may be largely dependent on the sealing resistance. Several reports can be found in literature describing recordings of neuronal activity, however, the experiments were performed with a wide variety in electrode properties, neurons used, and measurement conditions. The sealing resistance, for instance, depends on the size of the neurons as well as on the size of the electrode. A diversity of shapes and a wide range of measured amplitudes (40–350 μV) of recorded neuronal activity have been reported. Consequently, the investigation described in the following sections was focused on the occurrence of activity of dielectrophoretically trapped neurons and the timing, i.e., the first day at which neurons showed activity and the recorded spike rate.

4.3.2 MEA for DEP Trapping and Recording Neuronal Activity

The electrode structures used so far for trapping neuronal cells were quadrupole electrode structures of large dimensions (inter-electrode distance of 100 μm) as compared to cell dimensions (cell diameter≈10 μm). Therefore, in order to position a neuron or a group of neurons on top of all electrode tips of the microelectrode array (inter-electrode distances of 70 μm), new electrode configurations and procedures have had to be developed. The results expounded in the previous sections can be used as guides in the development of new (miniaturized) electrode structures for trapping groups of neuronal cells on top of each electrode tip of the MEA. These include the field strengths and frequencies needed to trap a distinct group of cells while keeping them alive.

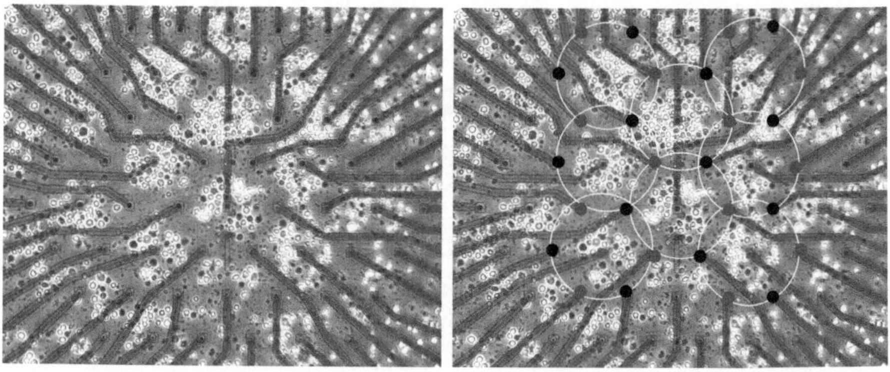

Fig. 43 Dissociated cortical rat neurons trapped on an MEA (**A**). Around each group of cells 6 electrodes were driven with a sinusoidal signal of 7 V_{PP} at 4 MHz (**B**)

In order to investigate neural activity of the trapped cells the MEA itself was used for the application of a nonuniform electrical field for the creation of the negative dielectrophoretic trapping forces. The result is shown in Fig. 43. The MEA contains 61 electrodes of which 28 were needed for the application of the nonuniform electric field in order to trap neurons, thus leaving, 33 electrodes available for recording neuronal activity of the trapped neurons.

The electrodes of the MEA are much smaller than the ones of the quadrupole electrode structure, while the distance between two opposing electrodes is greater (140 μm). The advantage of such small electrodes is that all cells are visible. In the case of the quadrupole structure a field minimum is also created on top of the electrodes where cells can be trapped without noticing them. These cells are not visible through the opaque gold electrodes using an inverted microscope.

The number of electrodes for the creation of the electric field can be reduced, but less compact groups will be trapped unless the field strength is increased. Finite element modeling has to be used to compute the exact field strengths to be created between four or six electrodes.

4.3.3 Experimental Procedure

4.3.3.1 Experimental Setup

For an optimal contact between neurons and electrodes used for recording purposes, the electrode impedance must be as low as possible; it was reduced by electrochemical deposition of a thin layer of platinum. Only those electrodes used for recording were platinized; the electrodes used for the creation of the nonuniform electric field were not.

The surface of the MEA inside the culture chamber was first coated with PEI, 30 ng/ml (Sigma-Aldrich), then the cortical cells were dissociated and the experi-

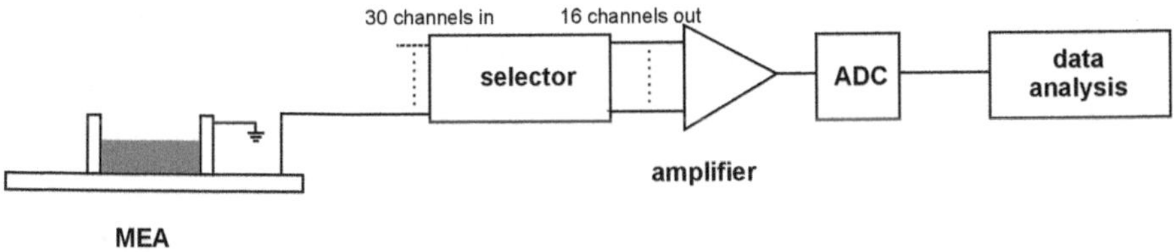

Fig. 44 Block diagram of the setup used for recording spontaneous neuronal activity from cortical rat cells

ments started. A drop of cell suspension (150 μl, 3×10^6 cells/ml) was pipetted on top of the electrode tips in the center of the MEA. In total 7 MEAs were used per experimental session of which 2 were used as a reference experiment, i.e., no electric field was applied. The culture chamber was sealed off whereafter the electrode plate was clamped into the small box in which small pins provide the connection to the signal generator. This box was directly placed on the microscope where the field was switched on (t=0). It was applied for 30 min (room temperature, minimal light exposure). Sinusoidal input signals of 7 and 8 V_{pp} at 1 and 4 MHz were used.

After 30 min the field was switched off and images were taken from the electrode tips by the digital video camera to observe trapping of the cells. After 30 min the electrode plate was removed from the microscope setup, taken out of the box and put away in the incubator (37°C, 5% CO_2, 100% humidity). After 2 h and 30 min, 2 ml of fresh R12 medium was applied to the culture chamber. Serum (10% FCS) was added to the medium, its addition being capable of substantially promoting the outgrowth of the neuronal network and growth of glial cells. It was found that a monolayer of neuronal cells was required in order to prevent death of cells not exposed fully to the nutrients of the medium when completely surrounded by other cells. Antibiotics (1% penicillin streptomycin) were also added to the medium to avoid infection (though knowing that these may negatively affect the spontaneous activity of the neurons). Medium was refreshed three times per week.

After about 7 DIV the neurons normally start to become active. Recording was started at 5 DIV and over a course of 6 days the MEAs were tested on active sites. When activity was found, data were recorded for a few minutes.

Measurements were performed according to the setup presented in Fig. 44. The amplifier consisted of 16 channels, so from 61 electrodes 16 had to be selected. From all electrodes recordings could be made due to the adaptability of this selection. The ring, which served as a culture chamber, was used as the contra electrode. The ring was coated with a thin platinum layer that provided the electrical contact with the medium.

The amplifier enlarged the recorded signals about 200 times after which the analogue signals were converted into digital signals. The sampling rate per channel was 12.5 kHz.

Signal recording and storage was performed using a program based on Labview (LabVIEW). In order to measure up to several minutes, data storage (400 kB/s) must be limited by rejecting data not containing bioelectrical activity. The standard deviation is measured or rms value of the signals of all 16 channels is determined when there is no activity. This noise level is multiplied by a user-defined value for the sig-

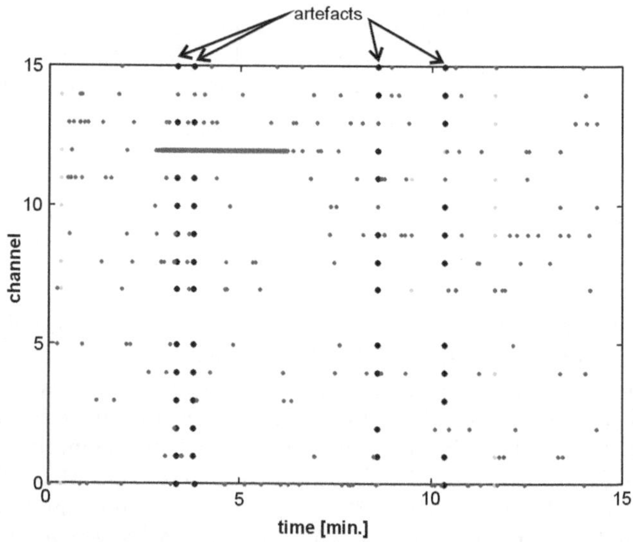

Fig. 45 Example of a period of 15 min of recording. Activity was clearly found on channel 12. Four artifacts are indicated; on several channels simultaneous activity was found

nal-to-noise ratio to form a threshold for activity detection. The data are analyzed in windows of 125 samples, i.e., 10 ms windows. When a sample in a window exceeds the threshold, this window is stored in a buffer together with the preceding and following window. Additional information, including the number of the channel at which the activity was detected, the starting time, and the noise level was also stored. After storage of 1,000 windows the buffer contents were written to a file.

4.3.3.2
Spike Analysis

The windows with action potentials were extracted from the data files by using Matlab scripts (Matlab 5.3, MathWorks). The first step of spike analysis was the removal of artifacts from biological activity. Externally generated noise signals, which may be picked up by the electrodes were very likely to occur simultaneously on several channels. In contrast, neurons are local sources of activity and therefore it is assumed that the chance that spikes are recorded on more than one electrode simultaneously is very small. Spikes occurring at the same time on two or more channels were regarded as artifacts and removed from the data for further analysis. Figure 45 shows an example of 15 min of recording with clear activity found on channel 12 while artifacts were detected on several channels. Assuming that the signals recorded from one electrode originate from one neuron, the spikes were averaged for each channel, thus reducing noise. In addition, for each channel the number of spikes per minute was determined.

Table 6 The number of MEAs on which activity was recorded

DIV	Reference	$7\ V_{PP}$		$8\ V_{PP}$
		1 MHz	4 MHz	4 MHz
5	0	0	0	0
6	0	0	1	0
7	2	0	2	1
8	5	0	5	1
9	3	0	4	1
10	1	0	2	0
Total	7	2	8	2

Table 7 Average spike rate (spikes/min)

	Reference	$7\ V_{PP}$
		4 MHz
Average	262	198
SD	139	30

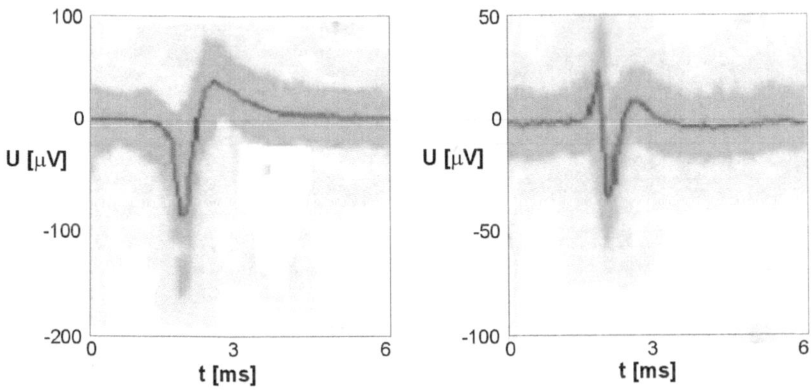

Fig. 46 Two averaged spike shapes recorded from neuronal cells trapped at $7\ V_{PP}$/4 MHz. Averages were taken from about 250 spikes for one channel [i.e., one (group of) neuron(s)]. From the reference cells similar spikes were found

4.3.4 Recording Results

Activity recordings were performed on a total of 20 MEAs for different field settings. Table 6 presents the number of MEAs on which active neurons were found and the number of days the cultures were in vitro.

The results found for the experiments performed at $7\ V_{PP}$/4 MHz showed results similar to the reference situation. Activity started after about 7 days. Dielectrophoretic trapping therefore does not seem to delay the appearance of neuronal activity, i.e., for cells trapped at 4 MHz. At 1 MHz no activity was observed, however, the number of experiments was low and it can therefore not be concluded yet that even though viability was preserved at 1 MHz, spontaneous activity was not present.

In Table 7 the average spike rate and standard deviation is presented for two situations. In the other situations the number of experiments was not sufficient for inclusion.

The spike rate in the reference situation varied between 112 and 511 spikes/min. while for the trapped situation (7 V_{pp}/4 MHz) this rate varied between 157 and 229 spikes/min. Variation was largest for the reference situation, but the spike rates were in the same range and can therefore be regarded as normal for the trapped cells.

Brief investigation of the shape of the spikes shows that the spikes have similar shapes. Two averaged shapes are shown in Fig. 46; these were found in the reference as well as in the trapped situation. Therefore, it might be concluded that cells trapped with fields created by sinusoidal signals of 7 and 8 V_{tt} at a frequency of 4 MHz did show normal activity after about 7 days; i.e., activity comparable to nonexposed cells. Further research may involve a more detailed investigation of the shape of the measured potentials, the firing pattern, and network properties of the trapped cells in comparison to reference cultures.

5 Summary

Trapping neuronal cells may aid in the creation of the cultured neuron probe. The aim of the development of this probe is the creation of the interface between neuronal cells or tissue in a (human) body and electrodes that can be used to stimulate nerves in the body by an external electrical signal in a very selective way. In this way, functions that were (partially) lost due to nervous system injury or disease may be restored. First, a direct contact between cultured neurons and electrodes is created. This is realized using a microelectrode array (MEA) which can be fabricated using standard photolithographic and etching methods. Section 1 gives an overview of the human nervous system, methods for functional recovery focused on the cultured neuron probe, and the prerequisites for culturing neurons on a microelectrode array.

An important aspect in the selective stimulation of neuronal cells is the positioning of cells or a small group of cells on top of each of the electrode sites of the MEA. One of the most efficient methods for trapping neuronal cells is to make use of dielectrophoresis (DEP). Dielectrophoretic forces are created when (polarizable) cells are located in nonuniform electric fields. Depending on the electrical properties of the cells and the suspending medium, the DEP force directs the cells towards the regions of high field strength (positive dielectrophoresis; PDEP) or towards regions of minimal field intensities (negative dielectrophoresis; NDEP). Since neurons require a physiological medium with a sufficient concentration of Na^+, the medium conductivity is rather high (~1.6 S/m). The result is that negative dielectrophoretic forces are created over the entire frequency range. With the use of a planar quadrupole electrode structure negative forces are directed so that in the center of this structure cells can be collected. The process of trapping cortical rat neurons is described in Sect. 2 theoretically and experimentally.

Medium and cell properties are frequency-dependent due to relaxation processes, which have a direct influence on the strength of the dielectrophoretic force. On the other hand, the nonideal material properties of the gold electrodes and glass substrate largely determine the electric field strength created inside the medium. Especially, the electrode–medium interface results in a significant loss of the input signal at lower frequencies (<1 MHz), and thus a reduction of the electric field strength inside the medium. Furthermore, due to the high medium conductivity, the electric field causes Joule heating. Local temperature rises result in local gradients in fluid density, which induces fluid flow. The electrode–medium interface and induced fluid flow are theoretically investigated with the use of modeling techniques such as finite element modeling. Experimental and theoretical results agreed with each other on the occurrence of the effects described in this section.

For the creation of the cultured neuron probe, preservation of cell viability during the trapping process is a prerequisite. Cell viability of dielectrophoretically trapped neurons has to be investigated. The membrane potential induced by the external field plays a crucial role in preservation of cell viability. The membrane can effectively be represented by a capacitance in parallel with a low conductance; with increasing frequency and/or decreasing field strength the induced membrane potential decreases. At high induced membrane potentials this representation for the membrane is no longer valid. At this point membrane breakdown occurs and the normally insulating membrane becomes conductive and permeable. The creation of electropores has been proposed in literature to be the cause of this high permeability state. Pores may grow or many small pores may be created which eventually may lead to membrane rupture, and thus cell death. Membrane breakdown may be reversible, but a chemical imbalance created during the high permeability state may still exist after the resealing of the membrane. This may cause cell death after several hours or even days after field application. Section 3 gives a detailed description of membrane breakdown. Since many investigations on electroporation of lipid bilayers and cell membranes are based on uniform electric fields, a finite element model is used to investigate induced membrane potentials in the nonuniform field created by the quadrupole electrode structure. Modeling results are presented in combination with the results of breakdown experiments using four frequencies in the range from 100 kHz to 1 MHz. Randomly positioned neuronal cells were exposed to stepwise increasing electric field strengths. The field strength at which membrane rupture occurred gives an indication of the maximum induced membrane potential. Due to the nonuniformity of the electric field, cell collapse was expected to be position-dependent. However, at 100 kHz all cells collapsed at a breakdown level of about 0.4 V, in contradistinction to findings at higher frequencies where more variation in breakdown levels were found. Model simulations were able to explain the experimental results.

For examining whether the neuronal cells trapped by dielectrophoresis were still viable after the trapping process, the frequency range was divided into two ranges. First, a high frequency (14 MHz) and a rather low signal amplitude (3 V_{pp}) were used to trap cells. At this high frequency the field-induced membrane potential is small according to the theoretical model, and therefore no real damage is expected. The experimental analysis included the investigation of the growth of the neurons, number and length of the processes (dendrites and axons), and the number of outgrowing (~viable) versus nonoutgrowing (~nonviable) neural cells. The experimental results agreed with the expectation.

The effect of the use of driving signals with lower frequencies and/or higher amplitudes on cell viability was investigated using a staining method as described in the second part of Sect. 4. Survival chances are not directly linked to the estimated maximum induced membrane potential. The frequency of the field plays an important role, decreasing frequency lowering the chance of survival. A lower frequency limit of 100 kHz is preferable at field strengths less than 80 kV/m, while with increasing field strength this limit shifts towards higher frequencies.

The theoretical and experimental results presented in this review form the inception of the development of new electrode structures for trapping neuronal cells on top of each of the electrodes of the MEA. New ways to investigate cell properties and the phenomenon of electroporation using electrokinetic methods were developed that can be exploited in future research linking cell biology to technology.

References

Asencor F J, Colom K, Santamaria C, Dominguez A, and Iglesias F J (1990). Comparison between the behavior of yeast and bacteria in a non-uniform alternating electric field. Bioelectrochem. Bioeng. 24, pp. 203–214

Awayda M S, Van Driessche W, and Helman S I (1999). Frequency-dependent capacitance of the apical membrane of frog skin: dielectric relaxation processes. Biophys. J. 76, pp. 219–232

Bank H L (1988). Rapid assessment of islet viability with Acridine Orange and Propidium Iodide. In Vitro Cellular and Developmental Biology, nr. 4, pp. 266–273

Banker G, and Goslim K. Culturing nerve cells, 2nd edition. A Bradford Book, The MIT Press, Cambridge, 1998

Bao J-Z, Davis C C, and Schmukler R E (1992). Frequency domain impedance measurements of erythrocytes: Constant phase angle impedance characteristics and a phase transition. Biophys. J. 61, pp. 1427–1434

Benz R, and Zimmermann U (1980a). Pulse length dependence of the electrical breakdown in lipid bilayer membranes. Biochim. Biophys. Acta 597, pp. 637–642

Benz R, Zimmermann U (1980b), The resealing process of lipid bilayers after reversible electrical breakdown. Biochim. Biophys. Acta 640, 169–178

Benz R, and Zimmermann U (1980c). Relaxation studies on cell membranes and lipid bilayers in the high electric field range. J. Electroanal. Chem. 116, pp. 723–739

Bove M, Grattarola M, Martinoia S, and Verreschi G (1995). Interfacing cultured neurons to planar substrate microelectrodes: characterisation of the neuron-to-microelectrode junction. Bioelectrochem. Bioeng. 38, pp. 255–256

Branch D W, Corey J .M, Weyhenmeyer J A, Brewer G J, and Wheeler B C (1998). Microstamp patterns of biomolecules for high-resolution neuronal networks. Med. Biol. Eng. Comp. 36, pp. 135–141

Buitenweg J R, Rutten W L C, Willems W P A, and Van Nieuwkasteele J W (1998). Measurement of sealing resistance of cell-electrode interfaces in neuronal cultures using impedance spectroscopy. Med. Biol. Eng. Comp. 36, pp. 630–637

Chang D C, Reese T S (1990). Changes in membrane structure induced by electroporation as revealed by rapid-freezing electron microscopy. Biophys. J. 58, pp. 1–12

Chiabrera A, Nicolini C, and Schwan H P. Interactions between Electromagnetic Fields and Cells. Plenum Press, New York, 1985

Cho M R, Thatte H S, Lee R C, and Golan D E (1996). Reorganization of microfilament structure induced by ac electric fields. FASEB J. 10, pp. 1552–1558

Clark P, Connolly P, Curtis A S G, Dow J.A T, and Wilkinson C D W (1987). Topographical control of cell behaviour: I. Simple step cues. Research Development Biology 99, pp. 439–448

Clark P, Connolly P, Curtis A S G, Dow J.A T, and Wilkinson C D W (1991a). Topographical control of cell behaviour: II. Multiple grooved substrata. Development 108, pp. 635–644

Clark P, Connolly P, Curtis A S G, Dow J.A T, and Wilkinson C D W (1991b). Cell guidance by ultrafine topography in vitro. J. Cell Sci. 99, pp. 73–77

Corey J.M, Wheeler B C, and Brewer G.J (1991). Compliance of hippocampal neurons to patterned substrate networks. J. Neurosci. Res. 30, pp. 300–307

Corey J.M, Wheeler B C, and Brewer G.J (1996). Micrometer resolution silane-based patterning of hippocampal neurons: Critical variables in photoresist and laser ablation processes for substrate fabrication. IEEE Trans. Biomed. Eng. 43, pp. 944–954
Coster H B L (1999). Self-assembly, stability and the electrical characteristics of cell membranes. Aust. J. Phys. 52, pp. 117–140
Crane J S, and Pohl H A (1968). A study of living and dead yeast cells using dielectrophoresis. J. Electrochem. Soc.: Electrochem. Sci. 115, pp.,584–586
Crane J S, and Pohl H A (1972). Theoretical models of cellular dielectrophoresis. J. Theor. Bio. 37, pp. 15–41
Curtis A, Wilkinson C, and Wojciak-Stothard B (1995). Cell guidance, movement and growth: Accelerating cell movement. Cell. Eng. 1, pp. 35–38
De Boer R W, and Van Oosterom A (1978). Electrical properties of platinum electrodes: impedance measurements and time-domain analysis. Med. Biol. Eng. Comput. 16, pp. 1–10
DeBruin K A, and Krassowska W (1999). Modeling electroporation in a single cell. I. Effects of field strength and rest potential, Biophys. J. 77, pp. 1213–1224
Drago G P and Ridella S (1982). Evaluation of electrical fields inside a biological structure. Br. J. Cancer 45, pp. 215–219
Dussaud A, Khusid B, and Acrivos A (2000). Particle segregation in suspensions subject to high-gradient ac electric fields. J. Appl. Phys. 88, pp. 5463–5473
Foster K R, Sauer F A, and Schwan H P (1992). Electrorotation and levitation of cells and colloidal particles. Biophys. J. 63, pp. 180–190
Fuhr G, Arnold W M, Hagedorn R, Müller T, Benecke W, Wagner B, and Zimmermann U (1992). Levitation, holding, and rotation of cells within traps made by high-frequency fields. Biochim. Biophys. Acta 1180, pp. 215–223
Fuhr G, Müller T, Schnelle T, Hagedorn R, Voigt A, Fiedler S, Arnold W M, Zimmermann U, Wagner B, and Heuberger A (1994a). Radio-frequency microtools for particle and live cell manipulation. Naturwissenschaften 81, pp. 528–535
Fuhr G, Fiedler S, Müller T, Schnelle Th, Glasser H, Lisec Th, and Wagner B (1994b). Particle micromanipulator consisting of two orthogonal channels with travelling-wave electrode structures. Sens. Act. A: Phys. 41–42, pp. 230–239
Fuhr G, Glasser H, Müller T, and Schnelle Th (1994c). Cell manipulation and cultivation under AC electric field influence in highly conductive culture media. Biochim. Biophys. Acta 1201, pp. 353–360
Fuhr G, and Shirley S G (1995). Cell handling and characterization using micron and submicron electrode arrays: state of the art and perspectives of semiconductor microtools. J. Micromech. Microeng. 5, pp. 77–85
Fuhr G, Voigt A, Müller T, Wagner B, Reimer K, and Lisec T (1995). Electric-field-mediated inhibition of cell and microparticle adhesion: a new way to create bio-repellent surfaces. Sens. Act. B: Chem. 26–27, pp. 468–470
Gascoyne P R C, Pethig R, Burt J P H, and Becker F F (1993). Membrane changes accompanying the induced differentiation of Friend murine erythroleukemia cells studies by dielectrophoresis. Biochim. Biophys. Acta 1149, pp. 119–126
Gascoyne P R C, Noshari J, Becker F F, and Pethig R (1994). Use of dielectrophoretic collection spectra for characterizing differences between normal and cancerous cells. IEEE Trans. Ind. Appl. 30, pp. 829–834
Gentet L J, Stuart G J, and Clemens J D (2000). Direct measurement of specific membrane capacitance in neurons. Biophys. J. 79, pp. 314–320
Gimsa J, Marszalek P, Loewe U, and Tsong T Y (1991). Dielectrophoresis and electrorotation of neurospora slime and murine myeloma cells. Biophys. J. 60, pp. 749–760
Gimsa J, and Wachner D (1998). A unified resistor-capacitor model for impedance, dielectrophoresis, electrorotation, and induced transmembrane potential. Biophys. J. 75, pp. 1107–1116
Glasser H, and Fuhr G (1998). Cultivation of cells under strong ac-electric field-differentiation between heating and trans-membrane potential effects. Biochem. and Bioenerg. 47, pp. 301–310
Glasser H, Schnelle Th, Müller T, and Fuhr G (1999). Electric field calibration in micro-electrode chambers by temperature measurements. Thermochimica Acta 333, pp. 183–190

Green N G, and Morgan H (1997). Dielectrophoretic separation of nano-particles. J. Phys. D: Appl. Phys. 30, pp. L41-L44
Green N G, and Morgan H (1997). Dielectrophoretic investigations of sub-micrometre latex spheres. J. of Phys. D: Appl. Phys., vol. 30, pp. 2626–2633
Gross G W (1979). Simultaneous single unit recording in vitro with a photoetched later deinsulated gold multimicroelectrode surface. IEEE Trans. Biomed. Eng. 26, pp. 273–279
Heida T, Rutten W L C, and Marani E (2001a). Dielectrophoretic trapping of dissociated fetal cortical rat neurons. IEEE Trans. Biomed. Eng. 48, pp. 921–930
Heida T, Rutten W L C, and Marani E (2001b). Viability of dielectrophoretically trapped neural cortical cells in culture. J. Neurosci. Meth. 110, pp. 37–44
Heida T., Rutten W.L.C. and Marani E. (2002a). Understanding dielectrophoretic trapping of neuronal cells: modeling electric field, electrode-liquid interface and fluid flow. J. Physics D: Appl. Phys. 35, pp. 1592–1602
Heida T., Rutten W.L.C. and Marani E. (2002b). Investigating membrane breakdown of neuronal cells exposed to nonuniform electric fields by finite element modeling and experiments. IEEE Trans. Biomed. Eng. 10, pp. 1195–1203
Ho S Y and Mittal G S (1996). Electroporation of cell membranes: a review. Crit. Rev. Biotech. 16, pp. 349–362
Huang Y, Hölzel R, Pethig R, and Wang X-B (1992). Differences in the AC electrodynamics of viable and nonviable yeast cells determined through combined dielectrophoresis and electrorotation studies. Phys. Med. Biol. 37, pp. 1499–1517
Hughes M P, Morgan H, Rixon F J, Burt J P H, and Pethig R (1998). Manipulation of herpes simplex virus type 1 by dielectrophoresis. Biochim. Biophys. Acta. 1425, pp. 119–126
Incropera F P, and De Witt D P. Fundamentals of heat and mass transfer. John Wiley and Sons, New York, Chichester, Brisbane, Toronto, Singapore, 1990
Irimajiri A (1979). A dielectric theory of "multi-stratified shell" model with its application to a lymphoma cell. J. Theor. Biol. 78, pp. 251–269
Irimajiri A, Asami K, Ichinowatari T, and Kinoshita Y (1987). Passive electrical properties of the membrane and cytoplasm of cultured rat basophil leukemia cells. I. Dielectric behavior of cell suspensions in 0.01–500 MHz and its simulation with a single-shell model. Biochimica et Biophysica Acta 896, pp. 203–213
Jones T B. Electromechanics of Particles, Cambridge University Press, New York, 1986
Jones T B, and Washizu M (1996). Multipolar dielectrophoretic and electrorotation theory. J. Electrostat. 37, pp. 121–134
Kaler K V I S, and Jones T B (1990). Dielectrophoretic spectra of single cells determined by feedback-controlled levitation. Biophys. J. 57, pp. 173–182
Kleinfeld D, Kahler K H, and Hockberger P E (1988). Controlled outgrowth of dissociated neurons on patterned substrates. J. Neurosci. 8, pp. 4098–4120
Kotnik T, Bobanović F, Miklavčič D (1997). Sensitivity of transmembrane voltage induced by applied electric fields – a theoretical analysis. Bioelectrochem. Bioenerg. 43, pp. 285–291
Kriegstein A R, and Dichter M A (1983). Morphological classification of rat cortical neurons in cell culture. J. Neurosci. 3, pp. 1634–1647
Lelong I H, Petgnief G, and Rebel G (1992). Neuronal cells mature faster on polyethylenimine coated plates than on polylysine coated plates. J. Neurosci. Res. 32, pp. 562–568
Levitan I B, and Kaczmarek L K. The neuron (Cell and molecular biology). Oxford university press, New York, 1991
Lide, D R (ed). CRC Handbook of Chemistry and Physics 74th edition, London: CRC, 1994
Marani E, Corino M, van den Berg R J, Rietveld W J, Deenen M, and Windhorst W (1988). Ionic conductances in cultured pre-infundibular cells from the hypothalamic arcuate region. Neuroendocrin. 48, pp. 445–452
Marha K, Musil J (1975). The cell as an electric circuit – I. Theoretical study. Biofizika 22, pp. 816–820
Markx G H, and Pethig R (1995). Dielectrophoretic Separation of Cells: Continuous Separation. Biotech. Bioeng. 45, pp. 337–343
Markx G H, Dyda P A, and Pethig R (1996). Dielectrophoretic separation of bacteria using a conductivity gradient. J. Biotechnol 51, pp. 175–180

Martini F H. Fundamentals of Anatomy and Physiology. Prentice Hall, New Jersey, 2001
Marszalek P, Liu D-S, and Tsong T Y (1990). Schwan equation and transmembrane potential induced by alternating electric field. Biophys. J. 58, pp. 1053–1058
Mason B D, and Townsley P.M (1971). Dielectrophoretic separation of living cells. Can. J. Microbiol. 17, pp. 879–888
Matsuzawa M, Potember R S, Stenger D A, and Krauthamer V (1993). Containment and growth of neuroblastoma cells on chemically patterned substrates. J. Neurosci. Meth. 50, pp. 253–260
Matsuzawa M, Liesi P, and Knoll W (1996). Chemically modifying glass surfaces to study substratum-guided neurite outgrowth in culture. J. Neurosci. Meth. 69, pp. 189–196
McAdams E T, Lackermeier A, McLaughlin J A, and Macken D (1995). The linear and non-linear electrical properties of the electrode-electrolyte interface. Biosens. Bioelec.10, pp. 67–74
Morgan H, Hughes M P, and Green N G (1999). Separation of submicron bioparticles by dielectrophoresis. Biophys. J. 77, pp. 516–525
Müller T, Gerardino A, Schnelle Th, Shirley S G, Bordoni F, De Gasperis G, Leoni R, and Fuhr G (1996). Trapping of micrometre and sub-micrometre particles by high-frequency electric fields and hydrodynamic forces. J. Phys. D: Appl. Phys. 29, pp. 340–349
Müller T, Gradl G, Howitz S, Shirley S, Schnelle Th, and Fuhr G (1999). A 3-D microelectrode system for handling and caging single cells and particles. Biosens. Bioelec. 14, pp. 247–256
Pastushenko V Ph, Kuzjmin P I, Chizmadhez Yu A (1985). Dielectrophoresis and electrorotation: A unified theory of spherically symmetrical cells. Studia Biophysica 110, pp. 51–57
Pethig R (1990). Application of AC electrical fields to the manipulation and characterisation of cells. Automation in Biotechnology—Proceedings on the 4th Conference—I.Karube (Ed.), pp. 159–185
Pethig R, Huang Y, Wang X-B, and Burt J P H (1992). Positive and negative dielectrophoretic collection of colloidal particles using interdigitated castellated microelectrodes. J. Phys. D: Appl. Phys. 24, pp. 881–888
Pine J (1980). Recording action potentials from cultured neurons with extracellular microcircuit electrodes. J. Neurosci. Meth. 2, pp. 19–31
Pohl H A. Dielectrophoresis. Cambridge University Press—Cambridge, 1978
Price J A R, Burt J P H, and Pethig R (1988). Applications of a new optical technique for measuring the dielectrophoretic behaviour of micro-organisms. Biochim. Biophys. Acta 964, pp. 221–230
Regehr W G, Pine J, Cohan C S, Mischke M D, and Tank D W (1989). Sealing cultured invertebrate cells to embedded dish electrodes facilitates long-term stimulation and recording. J. Neurosci. Meth.. 30, pp. 91–106
Romijn H J, van Huizen F, and Wolters P S (1984). Towards an improved serum-free, chemically defined medium for long-term culturing of cerebral cortex tissue. Neurosci. and Biobehav. Rev. 8, pp. 301–334
Ruardij T G, Goedbloed M H, and Rutten W L C (2000). Adhesion and patterning of cortical neurons on polyethylenimine and fluorocarbon coated surfaces. IEEE Trans. Biomed. Eng. 47, pp. 1593–1599
Rüegg U T, and Hefti F (1984). Growth of dissociated neurons in culture dishes coated with synthetic polymeric amines. Neurosci. Lett. 49, pp. 319–324
Rutten W, Mouveroux J-M, Buitenweg J, Heida C, Ruardij T, Marani E, and Lakke E (2001). Neuroelectronic interfacing with cultured multielectrode arrays toward a cultured probe. Proceedings IEEE, 89, pp. 1013–1029
Ryttsén F, Farre C, Brennan C, Weber S G, Nolkrantz K, Jardemark K, Chiu D T, and Orwar O (2000). Characterization of single-cell electroporation by using patch-clamp and fluorescence microscopy. Biophys. J. 79, pp. 1993–2001
Saito M, Schwan H P, and Schwarz G (1966). Response of nonspherical biological particles to alternating electric fields. Biophys.l J. 6, pp. 313–327
Schnelle Th, Hagedorn R, Fuhr G, Fiedler S, and Müller T (1993). Three-dimensional electric field traps for manipulation of cells—calculation and experimental verification. Biochim. Biophys. Acta 1157, pp. 127–140

Schnelle T, Müller T, Fiedler S, Shirley S G, Ludwig K, Herrmann A, Fuhr G, Wagner B, and Zimmermann U (1996). Trapping of viruses in high-frequency electric field cages. Naturwissenschaften 83, pp. 172–176

Schnelle T, Müller T, Fiedler S, and Fuhr G (1999). The influence of higher moments on particle behaviour in dielectrophoretic field cages. J. Electrostat. 46, pp. 13–28

Schwan H P. Electrical properties of tissue and cell suspensions. In: Advances in Biological and Medical Physics Vol. 5, Lawrence J.H., and Tobias E., Eds., Academic Press, New York, 1957

Schwan H P. Biophysics of the interaction of electromagnetic energy with cells and membranes. In: Biological effects and dosimetry of nonionizing radiation. Grandolfo M., Michaelson S.M., and Rindi A., eds. Plenum Press, New York, 1989

Shahar A, de Vellis J, Vernadakis A, Haber B. A dissection and tissue culture manual of the nervous system. Wiley-Liss, Inc., New York, 1989

Smit J P A, Rutten W L C, and Boom H B K (1999). Endoneural selective stimulation using wire-microelectrode arrays. IEEE Trans. Rehab. Eng. 7, pp. 399–412

Sukhorukov V L, Mussauer H, and Zimmermann U (1998). The effect of electrical deformation force on the electropermeabilization of erythrocyte membranes in low- and high-conductivity media. J. Membrane Biol. 163, pp. 235–245

Sukhorukov V L, Meedt G, Kürscher M, and Zimmermann U (2001). A single-shell model for biological cells extended to account for the dielectric anisotropy of the plasma membrane. J. Electrostat. 50, pp. 191–204

Sweeney J D, Ksienski D A, and Mortimer J T (1990). A nerve cuff technique for selective excitation of peripheral nerve trunk regions. IEEE Trans. Biomed. Eng. 37, pp. 706–715

Teissié J, and Rols M-P (1993). An experimental evaluation of the critical potential difference inducing cell membrane electropermeabilization. Biophys. J. 65, pp. 409–413

Thomas C A Jr., Springer P A, Loeb G E, Berwald-Netter Y, and Okun L M (1972). A miniature electrode array to monitor the bioelectrical activity of cultured cells. Exp. Cell Res. 74, pp. 61–66

Tsong T Y (1991). Electroporation of cell membranes. Biophys. J. 60, pp.297–306

Voldman J, Braff R A, Toner M, Gray M L, and Schmidt M A (2001). Holding forces of single-particle dielectrophoretic traps. Biophys. J. 80, pp. 531–541

Wang X-B, Pethig R, and Jones T B (1992). Relationship of dielectrophoretic and electrorotational behaviour exhibited by polarized particles. J. Phys. D: Appl. Phys. 25, pp. 905–912

Wang X-B, Huang Y, Hölzel R, Burt J P H, and Pethig R (1993). Theoretical and experimental investigations of the interdependence of the dielectric, dielectrophoretic and electrorotational behaviour of colloidal particles. J. Phys. D: Appl. Phys. 26, pp. 312–322

Wang X-B, Huang Y, Becker F F, and Gascoyne P R C (1994). A unified theory of dielectrophoresis and travelling wave dielectrophoresis. J. Phys. D: Appl. Phys. 27, pp. 1571–1574

Washizu M (1992). Precise calculation of dielectrophoretic force in arbitrary field. J. Electrostat. 29, pp. 177–188

Weaver J C (1993). Electroporation: a general phenomenon for manipulating cells and tissue. J. Cell. Biochem. 51, pp. 426–435

Weaver J C. Electroporation theory. From: Methods in Molecular Biology, vol. 47: Electroporation Protocols for Microorganisms, Humana Press Inc. Totowa, NJ, 2000

Wilkinson C D W (1995). Nanostructures in biology. Microelectron. Eng. 27, pp. 61–65

Zieliński J J, Marszalek P, and Fikus M (1989). A new method for the investigation of cellular dielectrophoresis. Z. Naturforsch. 44c, pp. 845–848

Zimmermann U, and Neil G A. Electromanipulation of cells. CRC Press, Boca Raton, Florida, 1996

Subject Index

GPSR Compliance

The European Union's (EU) General Product Safety Regulation (GPSR) is a set of rules that requires consumer products to be safe and our obligations to ensure this.

If you have any concerns about our products, you can contact us on ProductSafety@springernature.com

In case Publisher is established outside the EU, the EU authorized representative is:

Springer Nature Customer Service Center GmbH
Europaplatz 3
69115 Heidelberg, Germany

Batch number: 10704713

Printed by Printforce, the Netherlands